Soil–Machine Interactions

BOOKS IN SOILS, PLANTS, AND THE ENVIRONMENT

Soil Biochemistry, Volume 1, edited by A. D. McLaren and G. H. Peterson
Soil Biochemistry, Volume 2, edited by A. D. McLaren and J. Skujinš
Soil Biochemistry, Volume 3, edited by E. A. Paul and A. D. McLaren
Soil Biochemistry, Volume 4, edited by E. A. Paul and A. D. McLaren
Soil Biochemistry, Volume 5, edited by E. A. Paul and J. N. Ladd
Soil Biochemistry, Volume 6, edited by Jean-Marc Bollag and G. Stotzky
Soil Biochemistry, Volume 7, edited by G. Stotzky and Jean-Marc Bollag
Soil Biochemistry, Volume 8, edited by Jean-Marc Bollag and G. Stotzky
Soil Biochemistry, Volume 9, edited by G. Stotzky and Jean-Marc Bollag

Organic Chemicals in the Soil Environment, Volumes 1 and 2, edited by C. A. I. Goring and J. W. Hamaker
Humic Substances in the Environment, M. Schnitzer and S. U. Khan
Microbial Life in the Soil: An Introduction, T. Hattori
Principles of Soil Chemistry, Kim H. Tan
Soil Analysis: Instrumental Techniques and Related Procedures, edited by Keith A. Smith
Soil Reclamation Processes: Microbiological Analyses and Applications, edited by Robert L. Tate III and Donald A. Klein
Symbiotic Nitrogen Fixation Technology, edited by Gerald H. Elkan
SoilWater Interactions: Mechanisms and Applications, Shingo Iwata and Toshio Tabuchi with Benno P. Warkentin
Soil Analysis: Modern Instrumental Techniques, Second Edition, edited by Keith A. Smith
Soil Analysis: Physical Methods, edited by Keith A. Smith and Chris E. Mullins
Growth and Mineral Nutrition of Field Crops, N. K. Fageria, V. C. Baligar, and Charles Allan Jones
Semiarid Lands and Deserts: Soil Resource and Reclamation, edited by J. Skujinš
Plant Roots: The Hidden Half, edited by Yoav Waisel, Amram Eshel, and Uzi Kafkafi
Plant Biochemical Regulators, edited by Harold W. Gausman
Maximizing Crop Yields, N. K. Fageria

Modern Soil Microbiology, edited by J. D. van Elsas, J. T. Trevors, and E. M. H. Wellington

Growth and Mineral Nutrition of Field Crops: Second Edition, N. K. Fageria, V. C. Baligar, and Charles Allan Jones

Fungal Pathogenesis in Plants and Crops: Molecular Biology and Host Defense Mechanisms, P. Vidhyasekaran

Plant Pathogen Detection and Disease Diagnosis, P. Narayanasamy

Agricultural Systems Modeling and Simulation, edited by Robert M. Peart and R. Bruce Curry

Agricultural Biotechnology, edited by Arie Altman

Plant–Microbe Interactions and Biological Control, edited by Greg J. Boland and L. David Kuykendall

Handbook of Soil Conditioners: Substances That Enhance the Physical Properties of Soil, edited by Arthur Wallace and Richard E. Terry

Environmental Chemistry of Selenium, edited by William T. Frankenberger, Jr., and Richard A. Engberg

Principles of Soil Chemistry: Third Edition, Revised and Expanded, Kim H. Tan

Sulfur in the Environment, edited by Douglas G. Maynard

Soil–Machine Interactions: A Finite Element Perspective, edited by Jie Shen and Radhey Lal Kushwaha

Additional Volumes in Preparation

Mycotoxins in Agriculture and Food Safety, edited by K. K. Sinha and Deepak Bhatnagar

Soil–Machine Interactions

A Finite Element Perspective

Jie Shen

Radhey Lal Kushwaha

Department of Agricultural and Biosource Engineering
University of Saskatchewan
Saskatoon, Saskatchewan, Canada

MARCEL DEKKER, INC.   NEW YORK · BASEL · HONG KONG

Library of Congress Cataloging-in-Publication Data

Shen, Jie.
    Soil-machine interactions : a finite element perspective / Jie
Shen, Radhey Lal Kushwaha
      p.   cm. -- (Books in soils, plants, and the environment ; vol. 66)
      Includes bibliographical references and index.
      ISBN 0-8247-0145-3 (alk. paper)
      1.Soil-structure interaction--Mathmatical models. 2. Machinery,
Dynamics of--Mathematical models. 3. Finite element method.
    I. Kushwaha, Radhey Lal.   II. Title.   III. Series: Books in soils,
plants, and the environment ; v. 66.
    TA711.5.S48   1998
    624.1'5136 07--dc21   98-5188
                                                                        CIP

The publisher offers discounts on this book when ordered in bulk quantities. For more
information, write to Special Sales/Professional Marketing at the address below.

This book is printed on acid-free paper.

MARCEL DEKKER, INC.
270 Madison Avenue, New York, New York 10016
http://www.dekker.com

Current printing (last digit):
10 9 8 7 6 5 4 3 2 1

PRINTED IN THE UNITED STATES OF AMERICA

PREFACE

This book is an introduction to the finite element analyses of soil-machine systems, with a primary focus on soil-tool systems. Soil-machine systems designate systems in which the soil and tool or machine interact with each other for a specific purpose of production, such as tillage in agriculture, earth-moving in civil engineering, ditch-forming in military operation, hole-digging in forestry, tunnel-making in sea-bed operation, etc. Better tool performance in such systems will lead to a considerable saving in energy and labor, and an improvement in working efficiency in the daily operations mentioned above.

The study of soil-machine systems can be dated back to the early 1900s (Nichols, 1934; Goriatchkin, 1937). Since then, many investigations have been conducted using either an empirical or analytical approach, and can be generally categorized into:

1. Empirical determination of the parameters of soil-machine interaction;
2. Empirical determination of the performance of soil-machine systems using indoor or outdoor facilities;
3. Analytical determination of the interaction forces between soil and tool as well as the influence on the soil;
4. Analytical combination of the information of the interaction forces for parametric design of soil-engaging tools.

However, because of the complexity in soil-mechanical behavior, empirical results are hardly extended to a general case. The analytical approach (such as limit equilibrium method) cannot readily handle the effect of complex changes in three-dimensional geometry of tools on its performance.

In the last three decades, there has been a great expansion in the power and availability of numerical approaches, especially the finite element method (FEM). Yong and Hanna (1977) first applied the FEM in analyzing the soil-cutting process, opening a new era in the analysis and design of the tool or machine in a soil-machine system. The advantage of the FEM lies in its ability to deal with complex factors such as nonlinear mechanical behavior, dynamic loads and complex geometric configurations. Nowadays, the FEM analysis of the soil-machine system has been accepted and used in the research and teaching programs at various universities and colleges. However, no existing published reference book is available as an introduction to this new field.

The aim of this book is to present a unified treatment of finite element analysis in soil-machine systems and to combine the achievements and knowledge of many recognized investigators from different countries. It covers both theoretical and programming knowledge to implement the FEM in soil-machine systems and also demonstrates a number of case studies.

The potential benefits from using this book involve:

- Modernization of the design procedures for soil-engaging tools in agriculture, civil engineering, forestry, mining and military operations by using a rigorous and powerful method—the FEM.
- A potential savings of billions of dollars from these daily soil-processing practices if all soil-engaging tools are designed to operate efficiently.
- A good reference source for soil constitutive models and algorithms of the FE analyses.
- Assitance to users in the programming implementation of the FE analysis.

Since many excellent textbooks about the FEM are available on the market, the authors intentionally omitted some contents related to the basic theory and normal procedures of the FEM. However, the information on the FEM covered in this book is organized to be as self-contained as possible in terms of use.

The unique features of the book include:

- A thorough review of analytical methods in soil-tool interaction;
- Detailed introduction to soil constitutive models and soil-metal interface models;
- Systematic presentation of algorithms for static, dynamic, linear and non-linear FE analyses;
- A number of case studies which are very helpful for the reader's own implementations;
- Programming skills and tips on the FE analysis.

The first chapter contains background information on a number of soil-machine systems and an overview of a variety of recognized analytical methods. The deficiencies associated with those methods are given and one way to overcome them is discussed by means of the power of the finite element method. In Chapter 2, basic steps and formulation of the FEM are outlined and a simple example is given for a general view of what a finite element analysis is.

Accurate description of mechanical behavior of soils is essential for the success of a FE analysis of soil-tool system. Chapter 3 presents a comprehensive review of constitutive models of agricultural soils in three categories: elastic (linear, bi-linear, multi-linear and hyperbolic), plastic (rigid, perfectly plastic and elasto-plastic) and dynamic (visco-elastic, elasto-viscoplastic and rate-process).

The soil-metal (or soil-rubber) interface is a unique feature for a soil-machine system where adhesion and friction occur. To accurately predict the mechanical behavior of the interface, a special kind of interface element is needed for the FE analysis. Chapter 4 reviews typical interface elements developed in the past as well as the recognized models of stress-displacement relations. A case study of friction behavior at a soil-tool interface is given.

The basic computation of a FE analysis is a solving process of the equilibrium equations of the investigated system. Algorithms play a vital role in achieving better computation and/or storage efficiency of such a process. In Chapter 5, a variety of algorithms are introduced for static or dynamic, linear or non-linear cases, using a mixture of prose and FORTRAN keywords.

A number of distinguished studies have been conducted in FE analysis on soil-tool systems. Chapter 6 summarizes unique features of each of these studies, providing knowledge and experience to inform research and design activities in the future.

The FE analysis is a heavily computation-oriented process. As to the cases with the non-linear material and complex geometric configuration, computation or storage efficiency is still a major concern in the implementation of real analysis. A variety of techniques to improve time and/or space efficiency are introduced in Chapter 7. An introduction to computer memory hierarchy and algorithm timing analysis is beneficial for people in the engineering field to control and analyze the implementation of a FE analysis more easily.

The book has been arranged to be used as a text for senior undergraduate and graduate students, as a tool for designers, and as a reference for researchers in the area of soil dynamics, agricultural machinery, civil machinery, mining machinery, forestry machinery, military vehicle and space-explorer vehicles. By completing the study of the first five chapters, the student is assumed to have acquired the basic skills to conduct a FE analysis of a soil-tool system, as well as an understanding of soil constitutive models, interface element and equation solving algorithms. The tool or machine designer can use this book as a guide for the FE design activities of practical problems. The researcher can use it as a state-of-the-art reference for most of the recognized studies conducted in the past two decades.

We would like to acknowledge the financial support of Canadian Natural Science and Engineering Research Council and University of Saskatchewan. We are especially grateful to Dr. Liqun Chi, a former Ph.D. student in the Department of Agricultural and Bioresource Engineering, University of

Saskatchewan, who initiated the FEM soil-tool analysis. Some of those research results are included in this book. Final thanks to our wives, Lijia Zhu and K. Lata Kushwaha, for their patience and encouragement.

Jie Shen
Radhey Lal Kushwaha

CONTENTS

1

INTRODUCTION

With the continuous increase in computation speed and storage capacity of modern computers, the finite element method (FEM) has been feasibly applied in a variety of large-scale design or simulation problems. Even though it has originated from structural analyses, the theory upon which it is based is universally acceptable. The finite element method has successfully been used in simulating or solving many problems in other engineering fields including the soil-tool interaction in soil-machine systems.

1.1 STATEMENT OF THE PROBLEM

A soil-machine system in a broad sense might refer to any of the following systems:

1. *Soil-agricultural machinery*

The dynamic response of soil to farm tractors is one main factor in determining their overall tractive performance. The interaction between tillage tools and farm soil is of primary interest to the design and use of these tools for soil manipulation (McKyes, 1985).

2. *Soil-military machinery*

A variety of military tanks, tracks and other special-purpose vehicles operated on off-road ground are typical components in such a system (Bekker, 1956; Wong, 1993). Another instance consists of the working tools of machines for laying antitank mines or digging trenches, and the associated ground.

3. *Soil-forest machinery*

The success of a power-driven tree planter depends upon primarily the interaction between its working tool and the soil during the digging process. The soil compaction caused by machines in harvesting practices may have a detrimental effect on forest productivity (Grevers and Van Rees, 1995).

4. *Soil-mining and construction machinery*

Hydraulic excavators and bulldozers are widely accepted for quarrying, mining, earthmoving and construction applications (Claar et al., 1995). The performance and efficiency of these machines are dependent upon the interaction between soil and the working tools: shovel or blade.

In this book, only the cases of soil-tool interaction in soil-agricultural machinery systems are examined. However, the principles introduced here, with some minor modifications, can be applied to other cases mentioned above as well as the systems consisting of machines and other granular materials such as snow, peat, grains, ores, fertilizers, pesticides and soil amendments.

Without counting the approach of purely experimental evaluation, the methods studying the soil-tool interaction may be classified into two types:

- Analytical approach

The soil in front of a tool is broken up into several parts each of which is considered as a rigid object. The limit equilibrium method is applied to analyze the force balance in the entire system.

- Numerical approach

The soil-tool interaction is analyzed using numerical methods such as finite difference, finite element or boundary element. Since the finite element method has been the dominating numerical approach during the past twenty years in analyzing soil-tool interaction, this book will focus on how to apply it to the soil-tool interaction.

1.2 ANALYTICAL APPROACH—LIMIT EQUILIBRIUM

The limit equilibrium is one of the most important analytical approaches. The basic idea behind it is that soil and tool (or machine) are considered as a whole. The force equilibrium equations over the entire system are established with soil being in its limit state where its resistance becomes largest. From the equilibrium equations, forces acting on a tool or machine can be solved.

It should be kept in mind that the limit equilibrium method can be used to obtain only the information about the maximum forces which can be generated inside the soil body without providing much of a clue about how the soil body deforms. This is due to the fundamental assumptions embedded in the limit equilibrium. These assumptions make the method of a very simple form but quite limited in the power of analyzing the deformation in the system.

In general, the basic assumptions for a limit equilibrium method include:
(a) The soil is considered as a rigid material, that is, it is not deformable.
(b) The soil might fail inside the soil body and/or at a metal-soil interface. For the failure inside the soil body, one or several parts of soil may slide over a potential failure surface while with the failure at the interface, soil may slide over a metal-soil interface.
(c) The pattern of one or more failure surfaces inside the soil body is assumed or predetermined on an application-dependent basis. Actually, many patterns have been proposed by different investigators in the past. Each pattern may involve one or more unknown parameters leading to a series of potential failure surfaces. The unknown parameters are determined by performing an optimization to find the most critical failure surface which generates a minimum reaction force to a tool or machine.
(d) The forces interacting on a failure surface in the soil body are determined by the Mohr-Coulomb criterion:

$$\tau = c + \sigma_n \tan \phi$$

where: τ and σ_n = tangential and normal stresses, respectively
c = cohesion of soil
ϕ = internal frictional angle of soil

(e) The forces interacting on a metal-soil interface are determined by the following criterion:

$$\tau = a + \sigma_n \tan \delta$$

where: τ and σ_n = tangential and normal stresses, respectively
a = adhesion at a soil-tool interface
δ = external frictional angle at a soil-tool interface

On the basis of the above assumptions, the limit equilibrium method is ready to be applied. Two most important factors in this approach are:
(a) Shape of soil failure surfaces
    The shape is normally proposed on the basis of empirical observation or data, and is very crucial to the success in applying the limit equilibrium to an analysis of soil-machine systems.
(b) Equilibrium equations
    For two dimensional cases, the equilibrium equations in horizontal and vertical directions can be established by considering each individual soil block separated by failure surfaces in the soil body or metal-soil interfaces.

As to three dimensional cases, the equilibrium equations are set up in horizontal, side-way and vertical directions. By solving these equations, we can obtain the useful force information in a soil-machine system such as the draft or penetration force of a work tool.

In the past three decades, numerous numbers of studies have been conducted in applying the limit equilibrium in the soil-tool interaction. The more important ones are introduced below with a focus on the shape of soil failure surface and the form of equilibrium equations.

1.2.1 Two-Dimensional Models

The logarithmic spiral method, which was originally developed for the evaluation of soil loads in civil engineering (Terzaghi, 1959), has been used extensively in calculating soil resistance to tillage tools (Reece, 1965; Hettiaratchi and Reece, 1966, 1967, 1974), as shown in Fig. 1.1. The soil in front of a tool and above the failure surface is assumed to consist of two parts: (1) a Rankine passive zone (Prandtl, 1920) and (2) a complex shear zone bounded by part of a logarithmic spiral curve. The equilibrium equation of surface forces on the boundaries of these two parts as well as their body forces, can be used to calculate the horizontal and vertical forces on the tool.

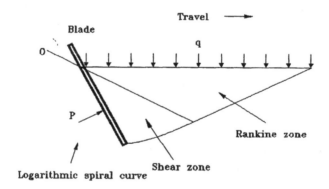

Figure 1.1 Logarithmic spiral failure zone.

Based upon the logarithmic spiral method, Reece proposed a simple earth-moving equation for the force necessary to cut soil with a tool as follows (Reece, 1965):

$$P = \gamma z^2 N_\gamma + czN_c + c_a zN_a + qzN_q \tag{1.1}$$

where: γ = specific weight of soil
c = soil cohesion
c_a = soil-metal adhesion
q = surcharge pressure

The four terms (N_γ, N_c, N_a and N_q) in equation (1.1) denote the gravitational, cohesive, adhesive and surcharge components of the soil reaction per unit width of the interface, respectively. These "N-factors" represent dimensionless numbers, which are functions of the geometry of soil-tool interfaces, soil international frictional angle, and soil-metal frictional angle (Hettiaratchi and Reece, 1966).

Fig. 1.1 and equation (1.1) are approximately valid for soil cutting tools with a width/depth ratio greater than unity. With narrow tillage tools, soil in front of a tool moves not only horizontally and vertically, but also sideways in the direction of the tool width. In this situation, the soil failure configuration becomes more complicated and the logarithmic spiral method is no longer sufficient in describing the three-dimensional failure surfaces.

1.2.2 Three-Dimensional Models

Based upon experimental observations and simplifications, several semi-empirical models have been proposed to evaluate forces required for the three-dimensional soil failure as follows.

Payne Model (1956) A three-dimensional soil failure model was evolved by Payne on the basis of the classical soil mechanics theories and a series of field and indoor tests on soil failure patterns. By observing the upward movement of soil in front of tools during tillage, a failure zone was postulated as shown in Fig. 1.2. This failure zone includes a triangular center wedge, a center crescent and two side blocks (called wings of the crescent). For tines with a width/depth ratio less than 1:1, the horizontal and vertical resultant forces on a tine are written as:

$$\begin{aligned} H = &B_c \cos\tau + B_R \sin(\phi + \tau) + T\cos\theta_m + \\ &+ 2[S_c \cos\beta \cos\lambda + S_R \cos\theta \sin(\alpha + \lambda)] \end{aligned} \tag{1.2a}$$

$$V = B_c \sin \tau - 2(S_c \sin \beta \cos \lambda + S_R \sin \theta) +$$
$$D_R \sin \delta - T \sin \theta_m + D_A + W \tag{1.2b}$$

where: B_C = cohesive force on the bottom surface of the center wedge

ϕ = internal frictional angle of soil

τ = inclination of the bottom surface of cleavage to the horizontal

B_R = resultant of the normal and frictional forces on the bottom of the wedge

T = force required to shear the front of the crescent

θ_m = inclination of T to the horizontal

S_C = cohesive force along one side of the wedge

α = angle between normal and projection of S_R in horizontal plane

β = inclination of S_C to the horizontal

λ = inclination to the direction of travel of the sides of the wedge

S_R = resultant of normal and frictional force on side of wedge

θ = angle between S_R and horizontal

α = angle between normal and projection of S_R in horizontal plane

D_A = adhesion of interface

W = weight of the soil in the wedge.

For the detailed formula and/or the procedures determining above variables, refer to (Payne, 1956).

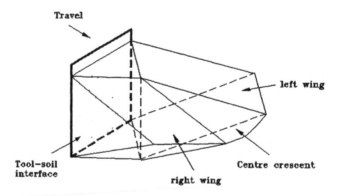

Figure 1.2 Failure zone of the Payne model.

One main shortcoming associated to the Payne model is that the entire procedure for solving the forces on a tool is quite complicated and time-consuming. Further extensive experiments showed that the shape of the failure zone was changed with the geometry of tools such as rake angle, depth and width (Payne and Tanner, 1959).

O'Callaghan-Farrelly Model (1964) Following Payne's work, O'Callaghan and Farrelly carried out a number of field tests with vertical plate tines on three different soils. Based on the field observations, a soil failure configuration was proposed as shown in Fig. 1.3. A critical depth is assumed to be 0.6× tine width. The upper portion of a tine corresponding to the depth less than the critical depth is viewed as a retaining wall. The failure surfaces caused by this portion are described by the two-dimensional logarithmic spiral method and the forces on two side surfaces are neglected. The lower portion of the tine is considered as a footing; the corresponding failure surfaces are described by the Prandtl rupture.

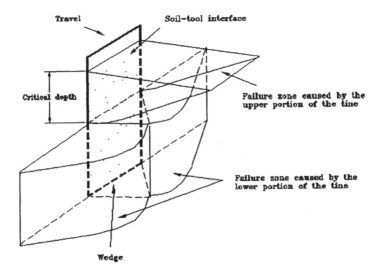

Figure 1.3 Failure zone of the O'Callaghan-Farrelly model.

For a shallow tine operating above the critical depth, a draft equation is developed by applying Terzaghi's method (Terzaghi, 1959) as follows:

$$H_s = w(czN_c + \gamma z^2 N_\gamma) \tag{1.3a}$$

The total draft of a deep tine, H_d, is given by adding the H_s in equation (1.3a) and the draft on the lower portion of the tine calculated by the Prandtl analysis.

$$H_d = \frac{cw(z-kw)}{\tan\phi}\left[\tan^2\left(\frac{\pi}{4}+\frac{\phi}{2}\right)e^{\pi\tan\phi}-1\right]+H_s \qquad (1.3b)$$

where: w = width of a tine
         k = critical ratio of depth/width

The predictions obtained from equations (1.3a) and (1.3b) are generally close to the test data except a certain amount of under prediction with the hardest soil encountered. But, it should be noted that in their study all test tines were flat and operated at a rake angle of $90°$. The gravity term, the second term on the right side of equation (1.3a) was neglected because of the small masses of soil and the low values of ϕ used. In addition, equations (1.3a) and (1.3b) do not include the effect of adhesion and external friction between soil and tool surface.

Hettiaratchi-Reece Model (1967)   After developing the well-known two-dimensional soil failure equation (equation 1.1), Hettiaratchi and Reece proposed a three-dimensional soil failure configuration. This was divided into a forward failure, ahead of a soil-tool interface, and a transverse failure, the horizontal transverse movement of the soil away from the center line of the interface, as shown in Fig. 1.4.

The total force on a tool is determined as a resultant of the forces from the forward failure and transverse failure. The force contributed by the forward failure is determined simply by using the two-dimensional soil failure equation (equation 1.1).   The force required for transverse failure is similar to the expression derived by O'Callaghan and Farrelly (1964) except that a gravitational component is included. The expressions of forces on a tine can be summarized by:

$$P_f = (\gamma z^2 N_\gamma + czN_c + c_a zN_a + qzN_q)w \qquad (1.4a)$$

$$P_s = [\gamma(d+q/\gamma)^2 wN_{sy} + cwdN_{sc}]K_a \qquad (1.4b)$$

$$H = P_f \sin(\alpha+\delta) + P_s \sin\alpha + c_a z\cos\alpha \qquad (1.4c)$$

$$V = P_f \cos(\alpha+\delta) + P_s \cos\alpha + c_a z \qquad (1.4d)$$

where: P_f= forward failure force component
         P_s = sideways failure force component

H = draught component of the resultant forces on a tine or blade
V = lift component of the resultant forces on a tine or blade
z = depth of tine or blade
c = cohesion
w = width of tine or blade
d = effective depth of tine or blade
q = surcharge pressure
α = rake angle from horizontal

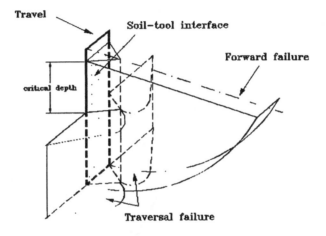

Figure 1.4 Failure zone of the Hettiaratchi-Reece model

The Hettiaratchi-Reece model includes the effects of soil properties, soil-metal properties and tool geometry such as rake angle, depth and width. However, the model is found to over-predict draft forces for vertical tools, while it under-predicts draft forces for inclined tools (Grisso and Perumpral, 1985).

Godwin-Spoor Model (1977) Godwin and Spoor investigated soil failure patterns with narrow tillage tines. Two separate failure regimes are assumed: (1) a three-dimensional crescent failure above a critical depth, and (2) a two-dimensional lateral (horizontal) failure below the critical depth. The configuration of the three-dimensional crescent soil failure is described using a parallel center wedge flanked with two curved side crescents, as shown in Fig. 1.5. The lateral failure below the critical depth is essentially similar to the horizontal failure proposed by O'Callaghan and Farrelly (1964) as well as Hettiaratchi and Reece (1967).

The total force on the upper failure surfaces is determined as a resultant of forces acting on all three sections in Fig. 1.5. Equation (1.1) and *N*-factors for soil failure are used to calculate the force on the linear section immediately ahead of the tine with its width equal to the tine width. The passive force *dP* of a crescent element acting at an angle η to the direction of travel, as shown in Fig. 1.5, is determined by the following equation:

$$dP = (\gamma d^2 N_\gamma + cd\ N_c + qd\ N_q)\frac{\gamma z \eta}{2} \qquad (1.5a)$$

where: η = angle of crescent element from the direction of travel
        d = critical depth
        z = depth of tine

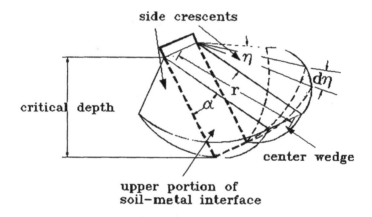

Figure 1.5 The upper failure zone of the Godwin-Spoor model. (From *Journal of Agricultural Engineering Research*, R.J. Godwin, et al., 1977. Reproduced with permission of Silsoe Research Institute, West Park, Silsoe, Bedford, U.K.)

An integration method can be applied to evaluate the total force on the side crescents. To simplify the integration, the failure boundary on the top surface is assumed to be circular. The draft and vertical forces for three-dimensional crescent failure above critical depth are given by:

$$H_t = (\gamma d^2 N_\gamma + cd\ N_c + qd\ N_q)(w + r\sin\eta)\sin(\alpha + \delta)$$
$$+ c_a wd\ [N_a \sin(\alpha + \delta) + \cos\alpha] \tag{1.5b}$$

$$V_t = -(\gamma d^2 N_\gamma + cd\ N_c + qd\ N_q)(w + r\eta\frac{\pi}{180})\cos(\alpha + \delta)$$
$$- c_a wd\ [N_a \cos(\alpha + \delta) + \sin\alpha] \tag{1.5c}$$

where: w = width of tine

$\quad r$ = crescent radius

$\quad \alpha$ = rake angle from the forward horizontal

$\quad q$ = surcharge pressure

$\quad \eta$ = an angle given by:

$$\eta = \cos^{-1}\left(\frac{d\ \cot\alpha}{r}\right) \tag{1.5d}$$

For the lateral soil failure, the total force H_l on the tine face below the critical depth is given by:

$$H_l = wcN_c'(z - d) + 0.5(1 - \sin\phi)\gamma w N_\gamma'(z^2 - d^2)$$
$$N_c' = \cot\phi\left[\frac{(1 + \sin\phi)e^{2\theta\tan\phi}}{(1 - \sin\phi\sin(2\eta + \phi))} - 1\right] \tag{1.5e}$$
$$N_q' = \frac{(1 + \sin\phi)e^{2\theta\tan\phi}}{1 - \sin\phi\sin(2\eta + \phi)}$$

The total draft force for a deep blade is, therefore, given as:

$$H = H_t + H_l \tag{1.5f}$$

Use of the Godwin-Spoor model requires prior knowledge of the rupture distance r (Fig. 1.5). Godwin and Spoor (1977) developed a graph using the information from Payne (1956), Payne and Tanner (1959), and Hettiaratchi and Reece (1967) to describe the relationship between distance ratio (rupture distance/depth) and tool angle. However, the determination of the rupture distance r is generally difficult.

McKyes-Ali Model (1977) In the McKyes-Ali model, a failure wedge is also proposed ahead of a cutting blade, as shown in Fig. 1.6. Similar to the

Godwin-Spoor model, the failure wedge consists of a center wedge and two side crescents. The only difference in failure shape is that the bottom surface of the center wedge is assumed as a plane and the bottom of the side crescent is assumed to be a straight line.

In the McKyes-Ali model, the forces on each section are determined by applying the mechanics of equilibrium directly rather than using equation (1.1) and N-factors of two-dimensional soil failure. A flat bottom plane of the center wedge and the straight line at the bottom of the crescents enable us to define the direction of reaction forces at the bottom of the failure zone. Forces contributed by the center wedge and side wedges are also considered. The proposed draft equation is similar to equation (1.1). However, the N-factors are re-evaluated for three-dimensional soil failure, which are given by the following equations:

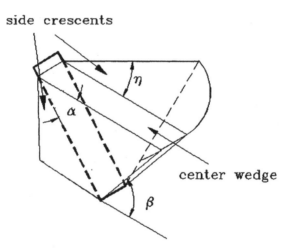

Figure 1.6   Single wedge failure zone of the McKyes-Ali model. (From *Journal of Terramechanics*, E. McKyes, et al., 1977. Reproduced with permission of the international society for Terrain Vehicle Systems, c/o U.S.A. CRREL, 72 Lyme Road, Hanover, NH.)

$$N_{\gamma H} = \frac{\dfrac{r}{2z}\left[1 + \dfrac{2r}{3w}\sin\eta\right]}{\cot(\alpha + \delta) + \cot(\beta + \phi)} \qquad (1.6a)$$

$$N_{cH} = \frac{[1 + \cot\beta \cos(\beta + \phi)][1 + \dfrac{r}{w}\sin\eta]}{\cot(\alpha + \delta) + \cot(\beta + \phi)} \qquad (1.6b)$$

$$N_{qH} = \frac{\dfrac{r}{w}\left[1 + \dfrac{r}{w}\sin\eta\right]}{\cot(\alpha + \delta) + \cot(\beta + \phi)} \qquad (1.6c)$$

where: z = operating depth
   r = rupture distance which is given as:

$$r = z(\cot\beta + \cot\alpha) \qquad (1.6d)$$

Each of the dimensionless factors (equation (1.6a) to equation (1.6c)) is a function of angle ß, as shown in Fig. 1.6. Angle ß is determined by minimizing the factor $N_{\gamma H}$ (the factor for the gravity term in draft). Then, it is used to compute the remaining *N*-factors. Compared to the Godwin-Spoor model, the McKyes-Ali model is easier to use and does not require prior knowledge of rupture distance. McKyes (1985) published a set of charts to determine the *N*-factors in equations (1.6.a) through equation (1.6.c) for some values of internal friction angle ϕ and external friction angle δ. Alternatively, a simple program can be written based on equation (1.6a) through equation (1.6c) to evaluate the *N*-factors.

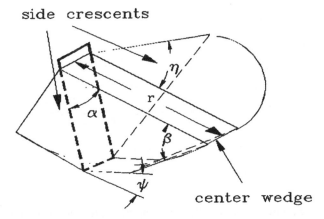

Figure 1.7   Double-wedge failure zone of the McKyes-Ali model. (From *Journal of Terramechanics*, E. McKyes, et al., 1977. Reproduced with permission of the international society for Terrain Vehicle Systems, c/o U.S.A. CRREL, 72 Lyme Road, Hanover, NH.)

McKyes and Ali (1977) compared the *N*-factors in equation (1.6a) through equation (1.6c) with the *N*-factors for two-dimensional soil cutting by setting $w = \infty$. The result shows a very good agreement for a smooth blade ($\delta = 0$). However, for a rough blade with a rake angle greater than $90 - \phi$, the *N*-factors in equation (1.6a) through equation (1.6c) are much higher than those for two-dimensional soil cutting. Therefore, a two-wedge model was proposed for a rough blade with a large rake angle, as shown in Fig. 1.7.

Perumpral-Grisso-Desai Model (1983) Perumpral et al. (1983) proposed another three-dimensional soil cutting model for narrow tillage tools. The model replaces the two side crescents flanking the center wedge by a set of two forces acting on the side of the center wedge as shown in Fig. 1.8. The authors claimed that the model had the same failure zone as the McKyes-Ali model. Since the soil weight of two side crescents is not considered and the side planes of center wedge are treated as slip plane, the actual failure zone of this model includes only a center wedge. Similar to the model by McKyes and Ali, the bottom slip surface is assumed to be straight. The model also assumes that the soil moves upward ahead of the tools (Fig. 1.8).

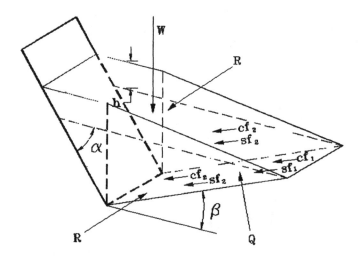

Figure 1.8 Failure zone of the Perumpral-Grisso-Desai model. (From *Transactions of the ASAE*, J. V. Perumpral, et al., 1983. Reproduced with permission of American Society of Agricultural Engineers, 2950 Niles Road, St. Joseph, MI.)

The total reaction force is determined by considering the equilibrium of all forces acting on the center wedge. The force equation is written into a form similar to equation (1.1):

$$P = w(\gamma z^2 N_\gamma + czN_c + c_a z N_a) \tag{1.7a}$$

and

$$N_\gamma = \frac{\dfrac{A}{wz^2}[2(1 - \sin\phi)z_a \sin\phi + w\sin(\phi + \beta)]}{\sin(\beta + \alpha + \phi + \delta)} \tag{1.7b}$$

$$N_c = \frac{\cos\phi\left[\dfrac{2A}{wz} + \dfrac{1}{\sin\beta}\right]}{\sin(\beta + \alpha + \phi + \delta)} \tag{1.7c}$$

$$N_a = \frac{-\left[1 + \dfrac{h}{z}\right]\cos(\beta + \alpha + \phi)}{\sin(\beta + \alpha + \phi + \delta)} \tag{1.7d}$$

where: z = tool depth
  h = height of soil heave in front of the tool at failure
  A = area of each side surface of the central wedge
  z_n = average depth at which the centroid of the failure wedge is located from soil surface:

$$z_a = \frac{1}{3}(z + h) \tag{1.7e}$$

The z_n in equation (1.7e) is a function of angle ß (Fig. 1.8) which is determined by minimizing the total force P. A computer program can be developed to minimize the force P and determine the failure plane angle ß.

Swick-Perumpral Model (1988) All the above models are static models in which the effect of travel speed is not considered. Swick and Perumpral proposed a dynamic soil cutting model that includes such an effect. The failure zone of the model, similar to the McKyes-Ali model, consists of a center wedge and two side crescents with a straight rupture plane at the bottom. In the

Godwin-Spoor and McKyes-Ali models, the extreme outer points of the side crescent are assumed to lie in a vertical plane passing through the forward tip of the tool. It was found that this assumption over-predicted the size of side crescents (Swick and Perumpral, 1988). Therefore, based on the observations from soil bin tests, the extended angle η is proposed as a function of rupture distance r and the rake angle α (Fig. 1.9) as:

$$\eta = \frac{\sin^{-1}(-6.03 + 0.46r + 0.0904\alpha)}{r} \tag{1.8a}$$

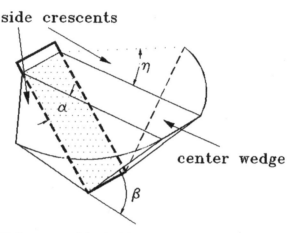

side crescents

center wedge

Figure 1.9 Failure zone of the Swick-Perumpral model.

The force equation of the model is derived in the same way as that for the McKyes-Ali model except that an acceleration force is added to account for the effect of travel speed. From the equilibrium equation of center wedge, the forces on center wedge are derived as:

$$P_1 = \frac{\dfrac{c_a z \cos(\alpha + \phi + \beta)}{\sin\alpha} + \left(\dfrac{\gamma\, zr}{2} + qr\right)\sin(\phi + \beta)}{\sin(\alpha + \phi + \beta + \delta)} + \frac{\left(\dfrac{cz}{\sin\phi} + f_{a1}\right)\cos\phi}{\sin(\alpha + \phi + \beta + \delta)} \tag{1.8b}$$

$$f_{a1} = \frac{\gamma \, zv^2 \sin\alpha}{g\sin(\alpha + \beta)} \qquad (1.8c)$$

where: g = gravitational constant, 9.8 m/s^2
        v = tool speed

The force contribution from the side crescents is derived by considering the equilibrium of a small slice of the side crescents and then integrating this force for whole side crescents. The tool force of one side crescent is obtained as:

$$P_2 = \frac{\left(\dfrac{\gamma \, zr^2}{6} + \dfrac{qr^2}{2}\right)\sin(\phi + \beta)\sin\eta}{\sin(\alpha + \phi + \beta + \delta)} \, w + \frac{f_{a2}\cos\phi\left(\dfrac{\eta}{2} + \dfrac{\sin 2\eta}{4}\right) + \dfrac{czr\cos\phi\sin\eta}{\sin\phi}}{\sin(\alpha + \phi + \beta + \delta)} \, w \qquad (1.8d)$$

$$f_{a2} = \frac{\gamma \, zrv^2 \sin\alpha}{2g\sin(\alpha + \beta)} \qquad (1.8e)$$

where: η = side crescents extended angle (Fig. 1.9) given by equation (1.8a).

The horizontal resultant force, P, on the center wedge and two side crescents is expressed as:

$$P = P_1 + 2P_2 \qquad (1.8f)$$

P is a function of rupture angle ß. According to the passive earth pressure theory, a passive failure takes place when the resistance from a soil wedge is minimum. The wedge creating a minimum resistance is found by minimizing P with respect to the rupture angle ß. A numerical procedure is not difficult to develop accordingly.

Zeng-Yao Model (1992) Zeng and Yao developed another dynamic soil cutting model which included the acceleration and strain-rate effect. This was obtained from the relation between soil shear strength and shear strain rate (Zeng and Yao, 1991), and the relation between soil-metal friction and sliding speed (Yao and Zeng, 1990). The failure zone of the model is similar to that of the McKyes-Ali model. One major difference between these two models is that the Zeng-Yao

model requires prior knowledge of failure shear strain for determining the
position of shear failure boundary, as shown in Fig. 1.10.

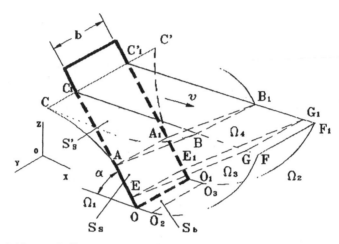

Figure 1.10   Failure zone of the Zeng-Yao model. (From *Journal of
Terramechanics*, D. Zeng, et al., 1992. Reproduced with permission of the
international society for Terrain Vehicle Systems, c/o U.S.A. CRREL, 72 Lyme
Road, Hanover, NH.)

In this model, the total draft P_X is divided into five components: compressive
force of soil along the blade P_G, side-edge shear force P_{SH}, inertia force of soil
in acceleration P_A, bottom-edge cutting force P_C and frictional force along the
cutting blade surface P_F, as in the following equation:

$$P_X = P_G \sin\alpha + (P_{SH} + P_A)\cos\beta + P_F \cos\alpha + P_C \tag{1.9a}$$

where: α = rake angle
        β = angle of absolute velocity of the soil slice with the ground surface

The total compressive force normal to the blade is:

$$P_G = 0.5b(z + L_2)(1 - \sin\phi)f\left[\frac{1 - \sin\beta}{\sin(\alpha + \beta)}\right] \tag{1.9b}$$

where: f = compressive stress, a function of strain ε, which can be determined with a compression test

L_2 = distance of the failure boundary from the cutting tip:

$$L_2 = \frac{z\varepsilon_f}{(1+\varepsilon_f)\sin\alpha} \tag{1.9c}$$

where: ε_f = failure shear strain

Applying the equation of soil shear strength as a function of shear rate and normal pressure, the side-edged shear force is expressed as:

$$P_{SH} = e^{C_1} L_2 \dot{\varepsilon}_f^{C_2} \frac{[1+C_4 f(1-\sin\phi)]^{C_3+1} - 1}{C_4 f(1+C_3)(1-\sin\phi)} \tag{1.9d}$$

where: C_1 to C_4 = soil parameters of dynamic shear strength

$\dot{\varepsilon}_f$ = shear strain rate at failure

Assuming a linear variation in speed and no direction change, the total inertia force is expressed as:

$$P_A = wz \frac{\gamma v^2 \sin\alpha}{g \sin^2(\alpha + \beta)} \tag{1.9e}$$

where: v = tool speed

The bottom edge cutting force and friction force along the cutting blade surface are expressed as:

$$P_C = wQ_0 t^{n-1} z^m \tag{1.9f}$$

$$P_F = c' + A' \ln\left[\frac{v\sin\beta}{\sin(\alpha+\beta)}\right] wz \tag{1.9g}$$
$$+ [(P_{SH} + P_A)\cos(90 - \alpha - \beta) + P_G]\tan\phi'$$

where: n, m = constants

t = thickness of the cutting edge

Q_0 = penetration parameter that varies with soil type, volume weight, moisture content and cutting speed

c', A', ϕ' = soil-metal dynamic friction parameters (for details, refer to Yao and Zeng (1990))

Although the above models do serve their purposes to a certain extent, the methodology used in developing these models is intrinsically of the following weaknesses:

(a) A failure profile is a prerequisite for the limit equilibrium analysis. However, the choice of the assumed profiles is arbitrary and depends on each particular investigator.

(b) Soil mechanical properties are assumed to be uniform without considering layered characteristic of some farm fields.

(c) The mode of soil failure is affected by the tool speed; and it is difficult to define such influence by tracing or describing the failure profile, a variant with the speed.

(d) Soil velocity and acceleration profiles in front of a tillage tool have to be simplified and assumed to follow a simple pattern without a sounding justification.

Evidently the above weaknesses may introduce errors in calculating forces in soil-tool interaction, especially in the cases different from the soil and tool conditions under which the model was developed. Hence, it is important to have a method that can alleviate these weaknesses to a certain extent.

1.3 NUMERICAL APPROACH —FINITE ELEMENT METHOD

The finite element method is a relatively new and effective numerical method, the advent of which is enhanced by the progress in modern computer architecture, programming languages and algorithms.

In the 1950s, the basic idea of the FEM originated from the matrix analysis of airplane structures in aviation engineering. According to the structure matrix analysis method, an entire structure can be considered as an assembly formed by linking many finite mechanical elements together; the function of each element is similar to the role of a brick for a building,

In 1960, this idea was extended in solving plane stress problems in elastic mechanics and a terminology of "finite element method" was adopted. However, for a continuum medium which is actually composed of infinite number of elements, the FEM can be used only after the continuum medium is discretized in the following manner:

(a) The continuum medium is divided into a finite number of blocks (or elements) which are linked to each other only at certain specified points, called nodes;

(b) Inside each element, the displacement distribution is approximated using a simple function and the relation between nodal force and nodal displacement is determined by the variation principle;

(c) Assemble the nodal force-displacement relation of all elements yielding a set of algebra equations with nodal displacements being unknowns. Solving such a set of equations provides the displacement information at a finite number of nodes within the continuum medium, i.e., the approximate solution to the problem.

Therefore, if the mechanical properties of each element are obtainable, the continuum medium can be analyzed by the FEM.

The FEM is different from classic analytical methods. In classic analytical methods, the mechanical properties of a micro-block are considered and a continuum medium is assumed to contain an infinite number of micro-blocks. By letting the size of the micro-block tend to be zero, a set of differential equations describing the mechanical properties of the continuum medium may be obtained. Solving such a set of equations leads to an analytical solution that provides the values of unknowns at any point within the continuum medium. However, as with most practical problems in engineering with irregular geometric shapes, non-linear and non-uniform material properties, an analytical solution is almost impossible. On the other hand, in the FEM, different elements can be assigned different material parameters to simulate the material non-uniformity. Iterative and incremental methods may be used to solve non-linear problems; arbitrary mesh generation eliminates the difficulty in modeling irregular geometric shapes. Thus, using the FEM may lead to an approximate solution to most problems in engineering.

The main advantages of the FEM include:

(a) It is suited for solving almost all the problems of continuum media and fields. It has been successfully used not only in stress analyses of non-uniform materials, anti-isotropic materials, non-linear stress-strain relations and complex boundary conditions, but also in heat transfer, fluid mechanics and electo-magnetics.

(b) It adopts the expression in matrix forms and thus is suited for computer programming. Once a general computer program is developed, it can be used for solving problems with any geometric shape simply by changing the input data.

The FEM can be implemented in three different ways: displacement-based, equilibrium and hybrid methods. In this book, we limit our discussions to the displacement-based finite element method which is a most important and widely-used formulation because of its simplicity, generality and good numerical properties.

2

BASIC PROCEDURES IN A FINITE ELEMENT ANALYSIS

The principles of the displacement-based finite element method for stress and strain analysis will be summarized in what follows (for more details, see Zienkiewicz, 1989).

2.1 BASIC STEPS IN A FINITE ELEMENT ANALYSIS

An analysis of the finite element method can be generally outlined into the following six steps:

1. Discretization of a system

The discretization of a system is the first step in a finite element analysis. It consists of the following: (1) the domain of a system is divided into a finite number of elements; (2) nodes are set up at specified points on each element; (3) adjacent elements are linked together only through these nodes; and (4) the assembly of all elements is used to replace the original system. If the analytical system is a structure consisting of only rods, such a division is obviously understandable because each rod can be defined as an element and rods actually interconnect each other only through their intersection points, i.e., nodes. However, as to a system with continuum medium as its domain, the shape, number and division pattern of elements should be paid more attention in order to effectively approximate it.

2. Determination of a displacement model

After the discretization of a system, the next step is to analyze the characteristics of a typical element. In order to express displacements, strains and stresses inside an element by using the nodal displacements, a displacement distribution within the element has to be assumed for a problem of continuum medium, that is, the displacement in the element is assumed to be a simple function of coordinates. This function is called a *displacement model* or *displacement function*.

The proper choice of a displacement function is a key step in an entire FEA. Most often, a displacement function is chosen as a polynomial because it provides some convenience with regard to mathematical operations such as derivation and integration, and it can locally approximate all continuous functions using an incomplete Taylor series. The maximum power and total number of terms in a polynomial are chosen on the consideration of the degrees of freedom of elements and the requirement for the solution convergence. Generally speaking, the total number of terms in a polynomial should be equal to the degrees of freedom of an element by counting both constant and linear terms.

On the basis of the chosen displacement function, displacements at any point within an element can be expressed by the nodal displacements in the following matrix notation:

$$\mathbf{u}_f = \mathbf{N}\,\mathbf{u} \tag{2.1}$$

where: $\mathbf{u}_f$ = the displacement column matrix for any point in an element
$\mathbf{u}$ = nodal displacement column matrix of an element
$\mathbf{N}$ = a shape function matrix, a function of coordinates

In the aspect of approximating the displacement distribution in a system, the FEM has obvious advantages over the conventional approximation methods. For example, in the classic Ritz method, a function is chosen to describe displacements in an entire system and to satisfy all the boundary conditions, while in the FEM an approximate displacement function is chosen with regard to an element (not the whole system) and such function needs not satisfy the boundary conditions of the entire system except for those at the interconnections with adjacent elements. Therefore, the implementation of the latter is much simpler than that of the former, especially in dealing with the problem of complex geometric configurations or complex materials. For a system with non-continuously distributed external loads, a piece-wise function is more suited than a continuous function in approximating the displacement function.

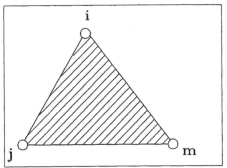

Figure 2.1 A triangular element.

• Displacement model of a constant-strain triangular element:

As a simple example, the displacement model for a triangular element shown in Fig. 2.1 may be defined as the following linear function:

$$\mathbf{u}_f = \begin{bmatrix} u \\ v \end{bmatrix} = \begin{bmatrix} N_i & 0 & N_j & 0 & N_m & 0 \\ 0 & N_i & 0 & N_j & 0 & N_m \end{bmatrix} \mathbf{u} \tag{2.2}$$

and

$$N_k = \frac{a_k + b_k x + c_k y}{2A} \qquad (k = i, j, m) \tag{2.3}$$

where: i, j and k = three nodes of a triangular element shown in Fig. 2.1

a_k, b_k and c_k = coefficients depending on x and y coordinates at node k ($k = i, j, m$), as expressed below:

$$a_i = x_j y_m - x_m y_j$$
$$b_i = y_j - y_m \tag{2.4}$$
$$c_i = -x_j + x_m$$

A = area of a triangular element given by:

$$A = \frac{1}{2} \begin{vmatrix} 1 & x_I & y_I \\ 1 & x_J & y_J \\ 1 & x_m & y_m \end{vmatrix} \tag{2.5}$$

$\mathbf{u}$ = element column matrix of nodal displacement defined by:

$$\mathbf{u} = \begin{bmatrix} u_i & v_I & u_J & v_J & u_m & v_m \end{bmatrix}^T \tag{2.6}$$

3. Analysis of mechanical properties of an element

After the displacement function is determined, it is ready for the analysis of mechanical properties of an element, which mainly includes the following three parts:

(a) On the basis of the geometric equation and equation (2.1), the relation between element strains and nodal displacements is derived as follows:

$$\varepsilon = \mathbf{B}\mathbf{u} \tag{2.7}$$

where: ε = the strain vector at any point in an element
$\mathbf{B}$ = the element strain matrix

(b) By combining the physical equation and equation (2.7), the relation between the element stresses and the nodal displacements can be expressed as:

$$\sigma = \mathbf{C}\mathbf{B}\mathbf{u} \tag{2.8}$$

where: σ = the stress vector at any point in an element
$\mathbf{C}$ = a constitutive matrix depending on the material properties of the element

(c) With reference to the virtual work principle and after some steps of derivations, the element stiffness equation, i.e. the relation between nodal force and nodal displacement on an element, is established as:

$$\mathbf{r} = \mathbf{k}\mathbf{u} \tag{2.9}$$

where: $\mathbf{k}$ = the element stiffness matrix which is determined by the virtual work principle as follows:

$$\mathbf{k} = \iiint_{vol} \mathbf{B}^T \, \mathbf{C} \, \mathbf{B} \, dx dy dz \tag{2.10}$$

in equation (2.10), *vol* signifies the 3-dimensional domain of an element. The element stiffness matrix is the kernel part of the analysis of element characteristics.

The typical formulas for a triangular element are listed below.

- Strain vector:

$$\varepsilon = \begin{bmatrix} \varepsilon_{xx} & \varepsilon_{yy} & \varepsilon_{xy} \end{bmatrix}^T$$

$$= \begin{bmatrix} \dfrac{\partial u}{\partial x} & \dfrac{\partial v}{\partial y} & \dfrac{\partial u}{\partial y} + \dfrac{\partial v}{\partial x} \end{bmatrix}^T \tag{2.11}$$

- The relation between strain vector and nodal displacement vector:

$$\varepsilon = \mathbf{B} \mathbf{u} \tag{2.12}$$

where: $\mathbf{B}$ = strain geometric matrix

$$\mathbf{B} = \begin{bmatrix} \mathbf{B}_i & \mathbf{B}_j & \mathbf{B}_m \end{bmatrix} \tag{2.13}$$

where:

$$\mathbf{B}_k = \begin{bmatrix} N_{k,x} & 0 \\ 0 & N_{k,y} \\ N_{k,y} & N_{k,x} \end{bmatrix} \qquad (k = i, j, m) \tag{2.14}$$

If N_k is determined by equation (2.2), then:

$$\mathbf{B}_k = \frac{1}{2A} \begin{bmatrix} b_k & 0 \\ 0 & c_k \\ c_k & b_k \end{bmatrix} \qquad (k = i, j, m) \tag{2.15}$$

- The relation between stress vector and nodal displacement vector:

$$\sigma = \begin{bmatrix} \sigma_{xx} \\ \sigma_{yy} \\ \sigma_{xy} \end{bmatrix} = \mathbf{C}\varepsilon = \mathbf{C}\mathbf{B}\mathbf{u} = \mathbf{S}\mathbf{u} \qquad (2.16)$$

where: $\mathbf{S}$ = stress matrix
$\mathbf{C}$ = elastic constitutive matrix

Stress matrix can be written in a more detailed form:

$$\mathbf{S} = \mathbf{C}\mathbf{B} = \begin{bmatrix} \mathbf{S}_i & \mathbf{S}_j & \mathbf{S}_m \end{bmatrix} \qquad (2.17)$$

where:

$$\mathbf{S}_k = \mathbf{C}\mathbf{B}_k \qquad (k = i, j, m) \qquad (2.18)$$

For elastic problems of plane stress or plane strain, C is of the form:

$$\mathbf{C} = \begin{bmatrix} C_1 & C_2 & 0 \\ C_2 & C_1 & 0 \\ 0 & 0 & C_3 \end{bmatrix} \qquad (2.19)$$

where:

$$C_1 = \frac{E[1 + (1 - flag)v]}{(1 + v)(1 - flag\ v)} \qquad (2.20)$$

$$C_2 = \frac{Ev}{(1 + v)(1 - flag\ v)} \qquad (2.21)$$

$$C_3 = \frac{E}{2(1 + v)} \qquad (2.22)$$

and

$$flag = \begin{cases} 2 & plane\ \ strain \\ 1 & plane\ \ stress \end{cases} \qquad (2.23)$$

where: E = elastic modulus
v = Poisson ratio

- Element static equilibrium equation:

$$\mathbf{k}\mathbf{u} = \mathbf{r} \tag{2.24}$$

where: $\mathbf{r}$ = element nodal force column matrix
$\mathbf{k}$ = element stiffness matrix

For a plane stress problem, element stiffness matrix $\mathbf{k}$ is of the following form:

$$\mathbf{k} = \begin{bmatrix} \mathbf{k}_{ii} & \mathbf{k}_{ij} & \mathbf{k}_{im} \\ \mathbf{k}_{ji} & \mathbf{k}_{jj} & \mathbf{k}_{jm} \\ \mathbf{k}_{mi} & \mathbf{k}_{mj} & \mathbf{k}_{mm} \end{bmatrix} \tag{2.25}$$

$$\mathbf{k}_{rs} = \frac{Ew}{4A(1-v^2)} \begin{bmatrix} b_r b_s + \dfrac{1-v}{2} c_r c_s & b_r c_s v + \dfrac{1-v}{2} c_r b_s \\ c_r b_s v + \dfrac{1-v}{2} b_r c_s & c_r c_s + \dfrac{1-v}{2} b_r b_s \end{bmatrix} \tag{2.26}$$

where: $r = i, j, m$
$s = i, j, m$

With a plane strain problem, the expression of $\mathbf{k}_{rs}$ in equation (2.26) should be modified by substituting $v/(1-v)$ with $E/(1-v^2)$.

4. Calculation of equivalent nodal forces

After a continuum medium is discretized, forces are assumed to pass from one element to another only through the nodes connecting these two elements. However, the forces in a continuum medium are actually transferred through the common boundary between two adjacent elements. Therefore, the surface forces acting on the boundary of an element, volumetric forces as well as concentrated forces on the element are all needed to be equivalently translated to forces acting on the nodes associated to that element. In other words, all forces acting on an element are replaced by their equivalent nodal forces. Such replacement is based upon the principle that the virtual work by two types of forces should be the same upon any virtual displacement.

5. Assembly of element stiffness equations and establishment of the equilibrium equations for an entire system

For a simple static problem, the establishment of equilibrium equations include both assembling element stiffness matrixes into the global stiffness matrix **K** and assembling the equivalent nodal force vectors of all elements into the global load column matrix **R**. The underlying reason for doing the assembly is that for all pairs of adjacent elements, the displacement at a common node is the same. After **K** and **R** are known, the static equilibrium equation for the entire system is:

$$\mathbf{KU} = \mathbf{R} \tag{2.27}$$

where:

$$\mathbf{K} = \sum_{i \;\in all\; elements} \mathbf{k}_i \tag{2.28}$$

$$\mathbf{R} = \mathbf{R}_1 + \mathbf{R}_2$$
$$= \mathbf{R}_1 + \sum_{i \;\in all\; elements} \mathbf{r}_{2i} \tag{2.29}$$

where: $\mathbf{R}_1$ = global concentrated load column matrix
$\mathbf{R}_2$ = global distributed load column matrix which may be either surface or body forces

Note that **K** in equation (2.27) is a singular matrix with its determinant being zero; only after the equation is revised by incorporating the geometric boundary conditions will **K** become non-singular and be used to solve all unknown nodal displacements.

6. Solution of unknown nodal displacements and calculation of element stresses

In a linear equilibrium problem in which the matrix **K** is considered constant, algorithms for solving linear equations can be used to solve the unknown displacement column matrix in equation (2.27). For a non-linear equilibrium problem with the matrix **K** being variant, an incremental procedure should be adopted to account for the gradual change of stiffness with stress states and paths.

Finally, the element stresses are calculated by using equation (2.8) and the nodal displacements.

- Magnitude and direction of principal stresses of a triangular element:

In two-dimensional cases, principal stresses of a triangular element is of the form:

$$\sigma_1 = \frac{\sigma_{xx} + \sigma_{yy}}{2} + \frac{(\sigma_{xx} - \sigma_{yy})^2}{\sqrt{2}} + \sigma_{xy}^2 \qquad (2.30)$$

$$\sigma_2 = \frac{\sigma_{xx} + \sigma_{yy}}{2} - \frac{(\sigma_{xx} - \sigma_{yy})^2}{\sqrt{2}} + \sigma_{xy}^2 \qquad (2.31)$$

where: σ_1, σ_2 = the major and minor principal stresses, respectively

The direction of principal stresses is expressed by:

$$\psi = \begin{cases} \arctan\left(\dfrac{\sigma_{xy}}{\sigma_{xx} - \sigma_2}\right) & \sigma_{xx} \neq \sigma_2 \\ 90° & \sigma_{xx} = \sigma_2 \end{cases} \qquad (2.32)$$

where: ψ = angle included between σ_1 and x axis

2.2 A SIMPLE EXAMPLE

Before discussing the problems of continuum media, let us study a simple structure with two articulated rods. This example is presented to understand the concept about how to determine mechanical properties of individual elements and how to obtain the force-displacement relation for the entire structure. Fig. 2.2 shows a structure consisting of two rods with a cross-sectional area A and an elastic modulus E. For the sake of simplicity, each rod can be viewed as an individual element with one node at each of its two ends. The lengths of these two rods, i.e. elements 1 and 2, are l_1 and l_2 respectively. Let Px_i and Py_i denote external forces at node i (=1, 2) in directions x and y, respectively. Fx_i^j and Fy_i^j represent internal nodal force components of element j (=1, 2) exerted by node i in directions x and y. u_i^j and v_i^j are displacement components, respectively in directions x and y, of node i associated to element j.

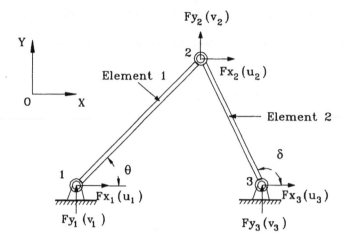

(a) Assembly of two rod elements

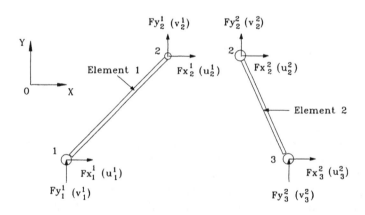

(b) the separated rod elements

Fig. 2.2 A structure with two articulated rods.

Since the rods are linked through hinges at which frictional moments can be neglected, the force and displacement at each node only have two components, i.e., two degrees of freedom. Consequently, each element in the structure has a total of four degrees of freedom. For element 1, the nodal force-displacement relation can be described by the following four equations:

$$
\begin{aligned}
Fx_1^1 &= k_{11}^1 u_1^1 + k_{12}^1 v_1^1 + k_{13}^1 u_2^1 + k_{14}^1 v_2^1 \\
Fy_1^1 &= k_{21}^1 u_1^1 + k_{22}^1 v_1^1 + k_{23}^1 u_2^1 + k_{24}^1 v_2^1 \\
Fx_2^1 &= k_{31}^1 u_1^1 + k_{32}^1 v_1^1 + k_{33}^1 u_2^1 + k_{34}^1 v_2^1 \\
Fy_2^1 &= k_{41}^1 u_1^1 + k_{42}^1 v_1^1 + k_{43}^1 u_2^1 + k_{44}^1 v_2^1
\end{aligned}
\tag{2.33}
$$

where, the superscripts refer to the element number. If the matrix notation is used, this equation is transformed to:

$$
\begin{bmatrix}
Fx_1^1 \\
Fy_1^1 \\
Fx_2^1 \\
Fy_2^1
\end{bmatrix}
=
\begin{bmatrix}
k_{11}^1 & k_{12}^1 & k_{13}^1 & k_{14}^1 \\
k_{21}^1 & k_{22}^1 & k_{23}^1 & k_{24}^1 \\
k_{31}^1 & k_{32}^1 & k_{33}^1 & k_{34}^1 \\
k_{41}^1 & k_{42}^1 & k_{43}^1 & k_{44}^1
\end{bmatrix}
\begin{bmatrix}
u_1^1 \\
v_1^1 \\
u_2^1 \\
v_2^1
\end{bmatrix}
\tag{2.34}
$$

or

$$
\mathbf{r}^1 = \mathbf{k}^1 \, \mathbf{u}^1
\tag{2.35}
$$

where: $\mathbf{u}^1 = \begin{bmatrix} u_1^1 & v_1^1 & u_2^1 & v_2^1 \end{bmatrix}^T$, nodal displacement vector of element 1

$\mathbf{r}^1 = \begin{bmatrix} r_{u1}^1 & r_{v1}^1 & r_{u2}^1 & r_{v2}^1 \end{bmatrix}^T$, nodal force vector of element 1

$\mathbf{k}^1$ = stiffness matrix containing stiffness coefficients k_{ij}^1

Similarly, the equilibrium equation for element 2 is expressed as:

$$
\begin{bmatrix}
Fx_2^2 \\
Fy_2^2 \\
Fx_3^2 \\
Fy_3^2
\end{bmatrix}
=
\begin{bmatrix}
k_{11}^2 & k_{12}^2 & k_{13}^2 & k_{14}^2 \\
k_{21}^2 & k_{22}^2 & k_{23}^2 & k_{24}^2 \\
k_{31}^2 & k_{32}^2 & k_{33}^2 & k_{34}^2 \\
k_{41}^2 & k_{42}^2 & k_{43}^2 & k_{44}^2
\end{bmatrix}
\begin{bmatrix}
u_2^2 \\
v_2^2 \\
u_3^2 \\
v_3^2
\end{bmatrix}
\tag{2.36}
$$

The stiffness coefficients, k_{ij}^1 and k_{ij}^2, in equations (2.34) and (2.36) can be determined using the virtual work principle. For instance, assume that element 1 is purely elastic and a set of virtual nodal displacements is given by:

$$u_1^1 = 1 \qquad v_1^1 = u_2^1 = v_2^1 = 0 \tag{2.37}$$

According to equation (2.34), the internal nodal forces of element 1 become:

$$Fx_1^1 = k_{11} \quad Fy_1^1 = k_{21} \quad Fx_2^1 = k_{31} \quad Fy_2^1 = k_{41} \tag{2.38}$$

Since the length of element 1 is decreased by $\Delta l_1 = \cos\theta$, as shown in Fig. 2.3, the required axial force is equal to:

$$\left(\frac{EA}{l_1}\right)\Delta l_1 = EA\frac{\cos\theta}{l_1} \tag{2.39}$$

which is the force exerted by node 1 on element 1. Its components in directions x and y are:

$$
\begin{aligned}
k_{11} &= \frac{EA}{l_1}\cos^2\theta \\
k_{21} &= \frac{EA}{l_1}\cos\theta\sin\theta
\end{aligned}
\tag{2.40}
$$

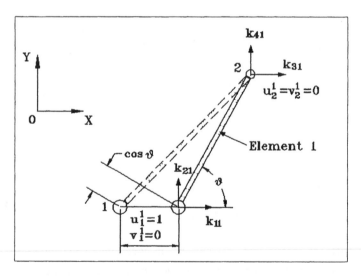

Figure 2.3 Element 1 at $u_1^1 = 1$, $v_1^1 = u_2^1 = v_2^1 = 0$.

As to the force exerted by node 2 on element 1, its magnitude is the same as above but its direction is opposite, i.e.,

$$
k_{31} = -\frac{EA}{l_1}\cos^2\theta
$$

$$
k_{41} = -\frac{EA}{l_1}\cos\theta\sin\theta
$$

(2.41)

Other entries in $\mathbf{k}^1$ and $\mathbf{k}^2$ can be obtained in a similar fashion leading to:

$$
\mathbf{k}^1 = \frac{EA}{l_1}
\begin{bmatrix}
c_\theta^2 & c_\theta s_\theta & -c_\theta^2 & -c_\theta s_\theta \\
c_\theta s_\theta & s_\theta^2 & -c_\theta s_\theta & -s_\theta^2 \\
-c_\theta^2 & -c_\theta s_\theta & c_\theta^2 & c_\theta s_\theta \\
-c_\theta s_\theta & -s_\theta^2 & c_\theta s_\theta & s_\theta^2
\end{bmatrix}
$$

(2.42)

$$
\mathbf{k}^2 = \frac{EA}{l_2}
\begin{bmatrix}
c_\phi^2 & -c_\phi s_\phi & -c_\phi^2 & c_\phi s_\phi \\
-c_\phi s_\phi & s_\phi^2 & c_\phi s_\phi & -s_\phi^2 \\
-c_\phi^2 & c_\phi s_\phi & c_\phi^2 & -c_\phi s_\phi \\
c_\phi s_\phi & -s_\phi^2 & -c_\phi s_\phi & s_\phi^2
\end{bmatrix}
$$

(2.43)

where:

$$
c_\theta = \cos\theta \qquad s_\theta = \sin\theta \qquad c_\phi = \cos\phi \qquad s_\phi = \sin\phi
$$

The assembly of nodal force-displacement relation of two elements can lead to the nodal force-displacement relation for the entire structure containing these two elements. Note that the relation between the global nodal displacement components of the structure, $u_1, v_1, u_2, v_2, u_3, v_3$, and the nodal displacement components of elements, u_1', v_1', u_2', v_2' (i denotes element number), is as follows:

$$
u_1 = u_1^1 \quad v_1 = v_1^1 \quad u_2 = u_2^1 = u_2^2
$$

$$
v_2 = v_2^1 = v_2^2 \quad u_3 = u_3^2 \quad v_3 = v_3^2
$$

(2.44)

The force equilibrium at each node requires that the external force on each node is equal to the sum of internal nodal forces on the elements which have a connection to the corresponding node, i.e.,

$$Px_1 = Fx_1^1 = \frac{EA}{l_1}(c_\theta^2\, u_1 + c_\theta s_\theta\, v_1 - c_\theta^2\, u_2 - c_\theta s_\theta\, v_2)$$

$$Py_1 = Fy_1^1 = \frac{EA}{l_1}(c_\theta s_\theta u_1 + s_\theta^2\, v_1 - c_\theta s_\theta\, u_2 - s_\theta^2\, v_2)$$

$$Px_2 = Fx_2^1 + Fx_2^2 = \frac{EA}{l_1}(-c_\theta^2\, u_1 - c_\theta s_\theta\, v_1 + c_\theta^2\, u_2 + c_\theta s_\theta\, v_2)$$

$$+ \frac{EA}{l_2}(c_\phi^2\, u_2 - c_\phi s_\phi\, v_2 - c_\phi^2\, u_3 + c_\phi s_\phi\, v_3)$$

$$Py_2 = Fy_2^1 + Fy_2^2 = \frac{EA}{l_1}(-c_\theta s_\theta\, u_1 - s_\theta^2\, v_1 + c_\theta s_\theta u_2 + s_\theta^2\, v_2)$$

$$+ \frac{EA}{l_2}(-c_\phi s_\phi u_2 + s_\phi^2 v_2 + c_\phi s_\phi\, u_3 - s_\phi^2 v_3)$$

$$Px_3 = Fx_3^2 = \frac{EA}{l_2}(-c_\phi^2\, u_2 + c_\phi s_\phi\, v_2 + c_\phi^2\, u_3 - c_\phi s_\phi\, v_3)$$

$$Py_3 = Fy_3^2 = \frac{EA}{l_2}(c_\phi s_\phi u_2 - s_\phi^2 v_2 - c_\phi s_\phi\, u_3 + s_\phi^2 v_3)$$

(2.45)

The above equation represents the nodal force-displacement relation for the entire structure. In matrix notation, it becomes:

$$
\begin{bmatrix} Px_1 \\ Py_1 \\ Px_2 \\ Py_2 \\ Px_3 \\ Py_3 \end{bmatrix} = EA
\begin{bmatrix}
p_1 & p_2 & -p_1 & -p_2 & 0 & 0 \\
p_2 & p_3 & -p_2 & -p_3 & 0 & 0 \\
-p_1 & -p_2 & p_1 + q_1 & p_2 - q_2 & -q_1 & q_2 \\
-p_2 & -p_3 & p_2 - q_2 & p_3 + q_3 & q_2 & -q_3 \\
0 & 0 & -q_1 & q_2 & q_1 & -q_2 \\
0 & 0 & q_2 & -q_3 & -q_2 & q_3
\end{bmatrix}
\begin{bmatrix} u_1 \\ v_1 \\ u_2 \\ v_2 \\ u_3 \\ v_3 \end{bmatrix}
$$

(2.46)

where:

$$p_1 = \frac{c_\theta^2}{l_1} \qquad p_2 = \frac{c_\theta s_\theta}{l_1} \qquad p_3 = \frac{s_\theta^2}{l_1}$$

$$q_1 = \frac{c_\phi^2}{l_2} \qquad q_2 = \frac{c_\phi s_\phi}{l_2} \qquad q_3 = \frac{s_\phi^2}{l_2}$$

Equation (2.46) can be written in another general form:

$$\mathbf{R} = \mathbf{KU} \qquad (2.47)$$

which is the basic equation to be established in the finite element method, in which $\mathbf{R}$ is the external load column matrix caused by external forces acting on the nodes; $\mathbf{U}$ is the column matrix composed of unknown nodal displacements; $\mathbf{K}$ is the global stiffness matrix of the structure. The assembly of stiffness coefficients associated to different elements can be illustrated more clearly by the following equation:

$$\mathbf{K} = \begin{bmatrix} k_{11}^1 & k_{12}^1 & k_{13}^1 & k_{14}^1 & 0 & 0 \\ k_{21}^1 & k_{22}^1 & k_{23}^1 & k_{24}^1 & 0 & 0 \\ k_{31}^1 & k_{32}^1 & k_{33}^1 + k_{11}^2 & k_{34}^1 + k_{12}^2 & k_{13}^2 & k_{14}^2 \\ k_{41}^1 & k_{42}^1 & k_{43}^1 + k_{21}^2 & k_{44}^1 + k_{22}^2 & k_{23}^2 & k_{24}^2 \\ 0 & 0 & k_{31}^2 & k_{32}^2 & k_{33}^2 & k_{34}^2 \\ 0 & 0 & k_{41}^2 & k_{42}^2 & k_{43}^2 & k_{44}^2 \end{bmatrix} \qquad (2.48)$$

The establishment of the global stiffness matrix is the kernel part in problem solving in the FEM. Once the global stiffness matrix is developed, the basic equation in the FEM can be listed without difficulty. The global stiffness matrix of the entire structure is formed by the superposition of stiffness matrixes of all elements. It can be seen, by observing equation (2.48), all terms with superscript 1 represent the stiffness coefficients of element 1, while the others with superscript 2 refer to the stiffness coefficients of element 2; the central part of $\mathbf{K}$ is generated by the summation of terms, from elements 1 and 2, destined at the same position in the global stiffness matrix. Therefore, the establishment of the global stiffness matrix requires the element stiffness matrixes to be determined first.

The global stiffness matrix has many characteristics. First, according to the Maxwell Commutative Theorem, it is a symmetric matrix. Secondly, all of the diagonal elements are positive, otherwise the direction of an acting force would be opposite to that of its corresponding displacement. The determinant corresponding to the matrix $\mathbf{K}$ is zero, i.e., it is singular. Therefore, equation (2.46) cannot be directly used to solve unknown nodal displacements. The physical reason for this is that no geometric constraints have been considered such that the rigid movement of the entire structure has not been removed. Only after the modification of the global stiffness matrix is conducted by eliminating the possibility of the rigid movement can equation (2.46) be used to solve all unknowns.

In the above, through a simple example, the physical sense and characteristics of the global stiffness matrix and its relationship with element stiffness matrixes are introduced. These characteristics are generally applicable and extendible to the problems in continuum media. However, for the complex continuum media such as soil, linear elasticity is no longer valid in many cases, leading to extra difficulty in determining the element stiffness matrices. The next chapter will introduce a variety of constitutive models suited for various soils.

3

CONSTITUTIVE MODELS FOR AGRICULTURAL SOILS

Constitutive models refer to the description of the relation between stress and strain with time being considered for dynamic cases. They are a central part of setting up a finite element analysis of a physical problem because they entirely determine the formulation of a constitutive matrix, $\mathbf{C}$, which is a kernel component of the basic FEM analytical equation as in equation (2.8).

Constitutive models can be classified into several categories on the basis of different criteria:

- *linear and non-linear models* — which depend on the linearity of equation.
- *elastic, plastic and elasto-plastic models* — which depend on whether only elasticity or plasticity or the both are considered in modeling or not.
- *static and dynamic (or rheological) models* — which depend on whether time is considered in modeling or not.

In this chapter, constitutive models for agricultural soils will be introduced in three categories: elastic, plastic and dynamic models. We begin with some basic concepts of stress and strain.

3.1 CONCEPTS OF STRESS AND STRAIN

3.1.1 Stress

Representation of a Stress State

Fig. 3.1 shows a three-dimensional state of stresses on an infinitesimal parallelepiped element of a larger continuous body, without counting the variation of stress with position inside this element. These stresses can be described in terms of either a stress tensor σ_{ij} $(i,j = x,y,z)$ or a matrix form shown by the following equation:

$$\sigma = \begin{bmatrix} \sigma_{xx} & \sigma_{xy} & \sigma_{xz} \\ \sigma_{xy} & \sigma_{yy} & \sigma_{yz} \\ \sigma_{xz} & \sigma_{yz} & \sigma_{zz} \end{bmatrix} \tag{3.1}$$

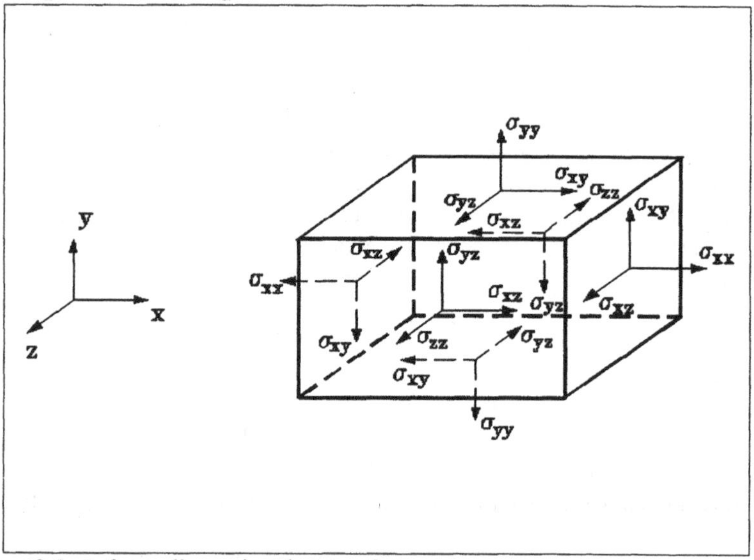

Figure 3.1 A three-dimensional state of stresses.

For the sake of clarity, a matrix form will be used in this book with a higher priority than a tensor notation.

Two-dimensional cases are of two classes: plane stress and plane strain. Suppose that we investigate stresses on an infinitesimal rectangular element in the x-y plane, as shown in Fig. 3.2. Plane stress means that there are no forces out of the plane, that is, σ_{zz}, σ_{xz} and σ_{yz} equal to zero; Plane strain means that no displacements occur out of the plane, that is, stresses σ_{xz}, σ_{yz} and strain ε_{zz} equal zero, but stress σ_{zz} does not vanish. Even though σ_{zz} is non-zero in plane strain cases, the condition $\varepsilon_{zz} = 0$ may be used to determine σ_{zz} based upon σ_{zz}, σ_{yy} and some other material parameters. Hence, we can process σ_{zz} separately and equation (3.1) shrinks into:

$$\sigma = \begin{bmatrix} \sigma_{xx} & \sigma_{xy} \\ \sigma_{xy} & \sigma_{yy} \end{bmatrix} \tag{3.2}$$

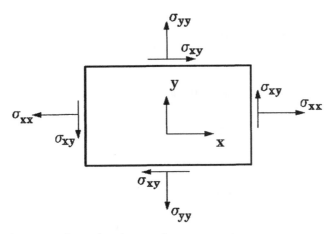

Figure 3.2 A two-dimensional state of stresses.

Principal Stresses
The stress σ in an original *x-y-z* coordinate system can be transformed into σ' in another *x'-y'-z'* system by

$$\sigma' = T \sigma T^T \tag{3.3}$$

$$T = \begin{bmatrix} \cos(x, x') & \cos(y, x') & \cos(z, x') \\ \cos(x, y') & \cos(y, y') & \cos(z, y') \\ \cos(x, z') & \cos(y, z') & \cos(z, z') \end{bmatrix} \tag{3.4}$$

where, **T** is the rotation matrix consisting of the direction cosines between two coordinate systems, *x-y-z* and *x'-y'-z'*. The angle between two arbitrary coordinates is represented by parentheses enclosing the corresponding coordinate labels separated by a comma, as shown in Fig. 3.3a.

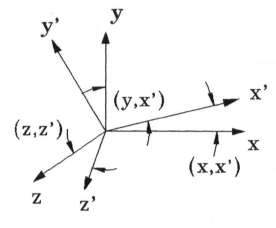

(a) from *x-y-z* to *x'-y'-z'*

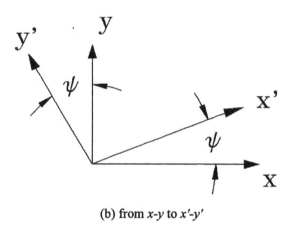

(b) from *x-y* to *x'-y'*

Figure 3.3 Rotation of coordinates.

According to the theory of matrices, a specific set of rotations of coordinates exists such that σ becomes:

$$\sigma' = \begin{bmatrix} \sigma_1 & 0 & 0 \\ 0 & \sigma_2 & 0 \\ 0 & 0 & \sigma_3 \end{bmatrix} \tag{3.5}$$

where: $\sigma_1, \sigma_2, \sigma_3$ = principal stresses which are the solutions of the following equations:

$$\sigma^3 - I_1\sigma^2 + I_2\sigma - I_3 = 0 \tag{3.6}$$

$$I_1 = \sigma_{xx} + \sigma_{yy} + \sigma_{zz} \tag{3.7}$$

$$I_2 = \sigma_{xx}\sigma_{yy} + \sigma_{xx}\sigma_{zz} + \sigma_{yy}\sigma_{zz} - \sigma_{xy}^2 - \sigma_{yz}^2 - \sigma_{xz}^2 \tag{3.8}$$

$$I_3 = \begin{vmatrix} \sigma_{xx} & \sigma_{xy} & \sigma_{xz} \\ \sigma_{xy} & \sigma_{yy} & \sigma_{yz} \\ \sigma_{xz} & \sigma_{yz} & \sigma_{zz} \end{vmatrix} \tag{3.9}$$

where: I_1, I_2 and I_3 = the first, second and third invariants of the stress, respectively.

For two-dimensional cases as shown in Fig. 3.3b, the rotation matrix **T** reduces to:

$$\mathbf{T} = \begin{bmatrix} \cos(x,x') & \cos(y,x') \\ \cos(x,y') & \cos(y,y') \end{bmatrix} = \begin{bmatrix} \cos\psi & \sin\psi \\ -\sin\psi & \cos\psi \end{bmatrix} \tag{3.10}$$

The magnitudes of principal stresses become:

$$\sigma_{1,2} = \frac{\sigma_{xx} + \sigma_{yy}}{2} \pm \sqrt{\left(\frac{\sigma_{xx} - \sigma_{yy}}{2}\right)^2 + \sigma_{xy}^2} \tag{3.11}$$

Octahedral Stresses
Octahedral stresses refer to the stresses on an octahedral plane which can be easily determined in the $\sigma_1 - \sigma_2 - \sigma_3$ coordinate system as in Fig. 3.4. They include octahedral normal stress σ_{oct} and octahedral shearing stress τ_{oct}. The orientations of σ_{oct} and τ_{oct} are, respectively, normal and tangential to an octahedral plane, while their magnitude are found from:

$$\sigma_{oct} = \frac{I_1}{3} = \frac{\sigma_1 + \sigma_2 + \sigma_3}{3} \tag{3.12}$$

$$\tau_{oct} = \frac{1}{3}\sqrt{(\sigma_1 - \sigma_2)^2 + (\sigma_2 - \sigma_3)^2 + (\sigma_3 - \sigma_1)^2} \tag{3.13}$$

$$\sigma_{oct} = \begin{bmatrix} \sigma_{oct} & 0 & 0 \\ 0 & \sigma_{oct} & 0 \\ 0 & 0 & \sigma_{oct} \end{bmatrix} \tag{3.14}$$

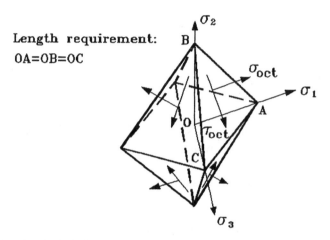

Length requirement:
OA=OB=OC

Figure 3.4 Octahedral planes and stresses.

Deviator Stresses

The deviator stresses are defined as:

$$\mathbf{s} = \begin{bmatrix} \sigma_{xx} - \sigma_{oct} & \sigma_{xy} & \sigma_{xz} \\ \sigma_{xy} & \sigma_{yy} - \sigma_{oct} & \sigma_{yz} \\ \sigma_{xz} & \sigma_{yz} & \sigma_{zz} - \sigma_{oct} \end{bmatrix} \tag{3.15}$$

Their principal stresses, s_1, s_2 and s_3 are determined by:

$$s^3 - J_2 s - J_3 = 0 \tag{3.16}$$

$$J_2 = \frac{1}{6}\left[(\sigma_{xx} - \sigma_{yy})^2 + (\sigma_{xx} - \sigma_{zz})^2 + (\sigma_{yy} - \sigma_{zz})^2 - \sigma_{xy}^2 - \sigma_{yz}^2 - \sigma_{xz}^2\right] \tag{3.17}$$

$$J_3 = \begin{vmatrix} \sigma_{xx} - \sigma_{oct} & \sigma_{xy} & \sigma_{xz} \\ \sigma_{xy} & \sigma_{yy} - \sigma_{oct} & \sigma_{yz} \\ \sigma_{xz} & \sigma_{yz} & \sigma_{zz} - \sigma_{oct} \end{vmatrix} \qquad (3.18)$$

3.1.2 Strain

Representation of a Strain State
The displacement of a body may be a consequence of deformation, rigid body motion or some combination of both. The deformation is described by the strain which is of two classes: engineering strain and tensorial shear strain. If the displacements in x, y and z directions are u, v and w, respectively, the engineering strains are defined as:

$$\varepsilon_{xx} = \frac{\partial u}{\partial x} \qquad \varepsilon_{yy} = \frac{\partial v}{\partial y} \qquad \varepsilon_{zz} = \frac{\partial w}{\partial z}$$

$$\gamma_{xy} = \frac{\partial u}{\partial y} + \frac{\partial v}{\partial x} \qquad \gamma_{xz} = \frac{\partial u}{\partial z} + \frac{\partial v}{\partial x} \qquad \gamma_{yz} = \frac{\partial v}{\partial z} + \frac{\partial w}{\partial y} \qquad (3.19)$$

A more succinct notation is tensorial shear strains defined by:

$$\varepsilon_{ij} = \frac{1}{2} \left(\frac{\partial u_i}{\partial x_j} + \frac{\partial u_j}{\partial x_i} \right) \qquad (i, j = x, y, z) \qquad (3.20)$$

$$\varepsilon = \begin{bmatrix} \varepsilon_{xx} & \varepsilon_{xy} & \varepsilon_{xz} \\ \varepsilon_{xy} & \varepsilon_{yy} & \varepsilon_{yz} \\ \varepsilon_{xz} & \varepsilon_{yz} & \varepsilon_{zz} \end{bmatrix} \qquad (3.21)$$

where, $u_x = u, u_y = v, x_x = x, x_y = y$, etc. The engineering strains are very useful for test data processing and numerical analysis, while the tensorial shear strains are more succinct and useful for theoretical derivations.

A two-dimensional state of strains is shown in Fig. 3.5 using an element with dimensions dx, dy and of unit thickness. Fig. 3.5a represents strains caused by a change in length of the sides, while Fig. 3.5b strains caused by a relative angle change without accompanying changes of length. Equations (3.19) and (3.20) reduce, respectively, to:

$$\varepsilon_{xx} = \frac{\partial u}{\partial x} \qquad \varepsilon_{yy} = \frac{\partial v}{\partial y} \qquad \gamma_{xy} = \frac{\partial u}{\partial y} + \frac{\partial v}{\partial x} \qquad (3.22)$$

$$\varepsilon_{xx} = \frac{\partial u}{\partial x} \qquad \varepsilon_{yy} = \frac{\partial v}{\partial y} \qquad \varepsilon_{xy} = \frac{1}{2}\left(\frac{\partial u}{\partial y} + \frac{\partial v}{\partial x}\right) \tag{3.23}$$

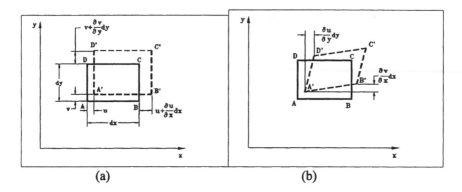

          (a)   (b)

Figure 3.5   A two-dimensional state of strain.

Principal Strains

If the notation of tensorial shear strains is used, the principal strains, $\varepsilon_1, \varepsilon_2$ and ε_3 are solutions to the equations:

$$\varepsilon^3 - K_1\varepsilon^2 + K_2\varepsilon - K_3 = 0 \tag{3.24}$$

$$K_1 = \varepsilon_{xx} + \varepsilon_{yy} + \varepsilon_{zz} \tag{3.25}$$

$$K_2 = \varepsilon_{xx}\varepsilon_{yy} + \varepsilon_{xx}\varepsilon_{zz} + \varepsilon_{yy}\varepsilon_{zz} - \varepsilon_{xy}^2 - \varepsilon_{yz}^2 - \varepsilon_{xz}^2 \tag{3.26}$$

$$K_3 = \begin{vmatrix} \varepsilon_{xx} & \varepsilon_{xy} & \varepsilon_{xz} \\ \varepsilon_{xy} & \varepsilon_{yy} & \varepsilon_{yz} \\ \varepsilon_{xz} & \varepsilon_{yz} & \varepsilon_{zz} \end{vmatrix} \tag{3.27}$$

where, K_1, K_2 and K_3 are called the first, second and third invariants of the strain, respectively.

Similarly, the magnitudes of principal strains in two-dimensional cases are:

$$\varepsilon_{1,2} = \frac{\varepsilon_{xx} + \varepsilon_{yy}}{2} \pm \sqrt{\left(\frac{\varepsilon_{xx} - \varepsilon_{yy}}{2}\right)^2 + \varepsilon_{xy}^2} \tag{3.28}$$

Octahedral Strains

An octahedral strain plane can be determined in the ε_1 - ε_2 - ε_3 coordinate system. The orientations of octahedral normal strain ε_{oct} and shear strain γ_{oct} are, respectively, normal and tangential to an octahedral plane, while their magnitudes are given by:

$$\varepsilon_{oct} = \frac{K_1}{3} = \frac{\varepsilon_1 + \varepsilon_2 + \varepsilon_3}{3} \tag{3.29}$$

$$\gamma_{oct} = \frac{1}{3}\sqrt{(\varepsilon_1 - \varepsilon_2)^2 + (\varepsilon_2 - \varepsilon_3)^2 + (\varepsilon_3 - \varepsilon_1)^2} \tag{3.30}$$

Deviator Strains

The definition of deviator strains is as follows:

$$\mathbf{e} = \begin{bmatrix} \varepsilon_{xx} - \varepsilon_{oct} & \varepsilon_{xy} & \varepsilon_{xz} \\ \varepsilon_{xy} & \varepsilon_{yy} - \varepsilon_{oct} & \varepsilon_{yz} \\ \varepsilon_{xz} & \varepsilon_{yz} & \varepsilon_{zz} - \varepsilon_{oct} \end{bmatrix} \tag{3.31}$$

Volumetric Strain

The volumetric strain is of the following form:

$$\varepsilon_{vol} = \frac{\Delta V}{V} = \varepsilon_{xx} + \varepsilon_{yy} + \varepsilon_{zz} = 3\varepsilon_{oct} \tag{3.32}$$

3.2 ELASTIC MODELS

The most distinguished feature of elastic models is that all strains are recovered when the load is removed. Within the scope of elasticity, models may be of linear or non-linear form as in Fig. 3.6. The latter can be further represented by bilinear, multi-linear and hyperbolic forms.

3.2.1 Linear Elastic Model

The stress-strain relation in a linear elastic model is linear as in Fig. 3.7a. If soil is considered to be completely isotropic, only two material parameters, shear modulus G and bulk modulus K, are required in the modeling as follows:

$$\sigma_{xy} = 2G\varepsilon_{xy} = G\gamma_{xy}$$
$$\sigma_{oct} = K\varepsilon_{vol}$$

(3.33)

In another way, a linear elastic model can be expressed in terms of Young's modulus E and Poisson's ratio v:

$$\sigma_{xx} = E\varepsilon_{xx}$$
$$\varepsilon_{yy} = \varepsilon_{zz} = -v\varepsilon_{xx}$$

(3.34)

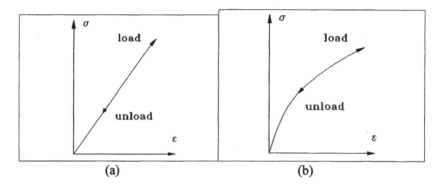

Figure 3.6   Elastic stress-strain models.

The following relations exist between the above two sets of material parameters:

$$G = \frac{E}{2(1+v)} \qquad K = \frac{E}{3(1-2v)}$$

(3.35)

The constitutive matrix for complete isotropic material is:

$$\mathbf{C} = \frac{E}{(1+v)(1-2v)} \begin{bmatrix} 1-v & v & v & 0 & 0 & 0 \\ v & 1-v & v & 0 & 0 & 0 \\ v & v & 1-v & 0 & 0 & 0 \\ 0 & 0 & 0 & 1-2v & 0 & 0 \\ 0 & 0 & 0 & 0 & 1-2v & 0 \\ 0 & 0 & 0 & 0 & 0 & 1-2v \end{bmatrix}$$

(3.36)

If soil is not considered completely isotropic, more material parameters are required to fill in the constitutive matrix. For example, in the situation of cross anisotropy where there is some plane in which stress-strain relations are isotropic and outside which the elastic constants for stresses and strains are different, the constitutive matrix needs five independent constants. If x-y plane is the plane of isotropy, the matrix is of the following form:

$$
\mathbf{C} = \begin{bmatrix}
C_{11} & C_{12} & C_{12} & 0 & 0 & 0 \\
C_{12} & C_{11} & C_{12} & 0 & 0 & 0 \\
C_{12} & C_{12} & C_{11} & 0 & 0 & 0 \\
0 & 0 & 0 & C_{11} - C_{12} & 0 & 0 \\
0 & 0 & 0 & 0 & C_{11} - C_{12} & 0 \\
0 & 0 & 0 & 0 & 0 & C_{11} - C_{12}
\end{bmatrix}
\tag{3.37}
$$

3.2.2 Bilinear and Multi-Linear Elastic Models

The simplest type of nonlinear relation is bilinear one as shown in Fig. 3.7a. Soil has an initial constitutive matrix $\mathbf{C}_1$ until the stresses reach a yield value σ_y, after which the constitutive matrix is changed to $\mathbf{C}_2$. The incremental stress-strain relation can be written as:

$$
\begin{aligned}
\Delta\sigma &= \mathbf{C}_1 \Delta\varepsilon & \sigma < \sigma_y \\
\Delta\sigma &= \mathbf{C}_2 \Delta\varepsilon & \sigma \geq \sigma_y
\end{aligned}
\tag{3.38}
$$

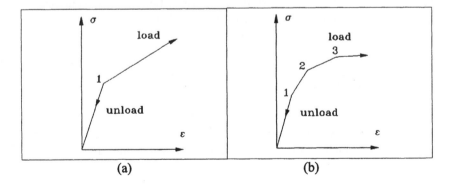

Figure 3.7 Bilinear and multi-linear elastic models.

If the constitutive matrix is expressed in terms of E and v, the Young's modulus is usually reduced and the Passion's ratio considered constant before and after the stresses reach σ_y. However, the drawback associated to this treatment is that the bulk modulus is reduced as much as the shear modulus. The soil element, therefore, becomes highly compressible and often an unreliable solution may follow. A better way to overcome this shortcoming is to express the constitutive matrix in terms of the shear modulus G and the bulk modulus K by reducing G and keeping K constant.

The dominant advantage of a bilinear model is its simplicity. However, the moduli E and G of real agricultural soils usually change gradually with the strains. To extend the bilinear models to accommodate this situation, multi-linear or piece-wise linear models are introduced as illustrated in Fig 3.7b. The tangent modulus E_t, which can be E or G, on a given n-piece-wised stress-strain curve is defined as the slope of the chord between two successive computed points as:

$$E_t = \frac{\sigma_i - \sigma_{i-1}}{\varepsilon_i - \varepsilon_{i-1}} \qquad 3 \leq i \leq n \tag{3.39}$$

This model gives a satisfactory description to a stress-strain curve. However, one unsatisfactory point of implementing this approach in that a cumbersome tabular procedure is required with a number of data points as input. In contrast, the following hyperbolic model only requires a few parameters to describe a stress-strain curve.

3.2.3 Hyperbolic Models

One of the most widely used constitutive models in finite element analyses is a hyperbolic one formalized by Duncan and Chang (1970), using a hyperbolic equation proposed by Kondner and his coworkers (1963, 1965):

$$(\sigma_1 - \sigma_3) = \frac{\varepsilon}{a + b\varepsilon} \tag{3.40}$$

Fig. 3.8a shows how the two parameters a and b define the shape of the stress-strain curve. These two parameters are determined by a transformed stress-strain curve in Fig. 3.8b on the basis of experimental data, and defined as:

$$a = \frac{1}{E_{si}} \tag{3.41}$$

$$b = \frac{1}{(\sigma_1 - \sigma_3)_{ult}} = \frac{R_{sf}}{(\sigma_1 - \sigma_3)_f} \tag{3.42}$$

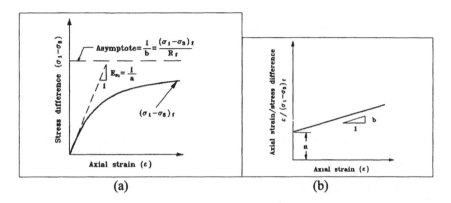

Figure 3.8 A hyperbolic stress-strain curve and its theoretically-transformed curve.

According to the experimental studies by Janbu (1963), E_{si} can be expressed as:

$$E_{si} = K_s\, pa \left(\frac{\sigma_3}{pa}\right)^{ns} \tag{3.43}$$

The Mohr-Coulomb failure criterion is expressed as:

$$(\sigma_1 - \sigma_3)_f = \frac{2c\cos\phi + 2\sigma_3 \sin\phi}{1 - \sin\phi} \tag{3.44}$$

Substitution of equations (3.41) through (3.44) into equation (3.40) yields the general form of the stress-strain relation in the Duncan-Chang model as follows:

$$(\sigma_1 - \sigma_3) = \frac{\varepsilon}{\left[K_s^{-1}\, pa^{-1}\left(\dfrac{\sigma_3}{p_a}\right)^{-ns} + \dfrac{\varepsilon\, R_{sf} f(1 - \sin\phi)}{2c\cos\phi + 2\sigma_3 \sin\phi} \right]} \tag{3.45}$$

For incremental stress analyses, the tangent modulus, E_S, is expressed as:

$$E_s = \frac{\partial(\sigma_1 - \sigma_3)}{\partial \varepsilon}$$
(3.46a)

By the substitution of equation (3.45) into equation (3.46a) and several derivations, equation (3.46a) can finally be transformed to:

$$E_s = \left[1 - \frac{R_s f(1 - \sin \phi)(\sigma_1 - \sigma_3)}{2c \cos \phi + 2\sigma_3 \sin \phi}\right]^2 K_s p_a \left(\frac{\sigma_3}{p_a}\right)^{n_s}$$
(3.46b)

In non-linear elastic cases, the constitutive matrix of one soil element is composed of the tangent modulus, E_S, and the Poisson's ratio, v_S, as follows:

$$C = \begin{bmatrix} C_{11} & C_{12} & C_{12} & 0 & 0 & 0 \\ C_{12} & C_{11} & C_{12} & 0 & 0 & 0 \\ C_{12} & C_{12} & C_{11} & 0 & 0 & 0 \\ 0 & 0 & 0 & C_{33} & 0 & 0 \\ 0 & 0 & 0 & 0 & C_{33} & 0 \\ 0 & 0 & 0 & 0 & 0 & C_{33} \end{bmatrix}$$
(3.47)

where:

$$C_{11} = \frac{E(1 - v)}{(1 + v)(1 - 2v)}$$

$$C_{12} = \frac{Ev}{(1 + v)(1 - 2v)}$$

$$C_{33} = \frac{E}{2(1 + v)}$$

In summary, equations (3.45) and (3.46b) are the ordinary and incremental forms of the Duncan-Chang model, respectively. Researchers usually adopted the incremental form in their non-linear analyses, because of the non-linearity of soils.

3.2.4 Rate-Dependent Hyperbolic Models

It was found by Yu and Shen (1988) that an approximate linear relation existed between stress and logarithm of strain rate in constant stress and constant strain rate tests of some agricultural soils. On the basis of experimental data from a triaxial apparatus, as shown in Figs. 3.9a and 3.9b, the shear difference of soil at failure can be approximately related to the strain rate in the following form:

$$(\sigma_1 - \sigma_3)_f = (\sigma_1 - \sigma_3)_{f0} + B_s \ln\left(\frac{\dot{\varepsilon}}{\dot{\varepsilon}_0}\right) \qquad \dot{\varepsilon} > \dot{\varepsilon}_0$$

$$(\sigma_1 - \sigma_3)_f = (\sigma_1 - \sigma_3)_{f0} \qquad \dot{\varepsilon} \le \dot{\varepsilon}_0$$

(3.48a)

where, $\dot{\varepsilon}_0$ is the base strain rate of soil specimens in the triaxial tests, a very small magnitude (0.00019 second^{-1}) in this study, below which the rate effect on the shear strength can be neglected.

If c and ϕ denote the Mohr-Coulomb parameters obtained at the base strain rate $\dot{\varepsilon}_0$, $(\sigma_1 - \sigma_3)_{f0}$ is determined by:

$$(\sigma_1 - \sigma_3)_{f0} = \frac{2c\cos\phi + 2\sigma_3 \sin\phi}{1 - \sin\phi}$$

(3.48b)

By the substitution of equations (3.48a) and (3.48b) into equation (3.46b), a modified strain-rate-dependent model for the tangential modulus is expressed as:

$\dot{\varepsilon} > \dot{\varepsilon}_0$:

$$E_s = K_s\, p_A \left(\frac{\sigma_3}{p_A}\right)^{n_s}\left[1 - \frac{R_{sf}(1-\sin\phi)(\sigma_1 - \sigma_3)}{2c\cos\phi + 2\sigma_3 \sin\phi + B_s(1-\sin\phi)\ln(\dot{\varepsilon}/\dot{\varepsilon}_0)}\right]^2$$

$\dot{\varepsilon} \le \dot{\varepsilon}_0$:

$$E_s = K_s\, p_a \left(\frac{\sigma_3}{p_a}\right)^{n_s}\left[1 - \frac{R_{sf}(1-\sin\phi)(\sigma_1 - \sigma_3)}{2c\cos\phi + 2\sigma_3 \sin\phi}\right]^2$$

(3.49)

3.3 PLASTIC MODELS

The theory of plasticity was originally developed on the basis of experiments on metals. Therefore, some basic concepts in soil plasticity were borrowed from those of metal plasticity. Fig. 3.10 shows a typical stress-strain relation for a metal bar. The initial state of the bar is supposed to be at point *O*. At the beginning portion of loading, designated by *OA,* the stress-strain relation is invertible, that is, loading is reversible. However, experiments revealed the existence of a certain point *B* beyond which the loading is irreversible. This point is called *yield point*. When a loading exceeds the yield point, any following unloading can only partially restore the developed total strain. The recovered and non-recovered component of the total strain are called *elastic strain* and *plastic strain*, respectively, as designated by *DE* and *OD* in the figure. The maximum stress that the material is able to sustain, is called *failure point*, as illustrated by point *F* in Fig. 3.10.

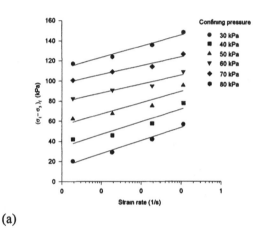

(a)

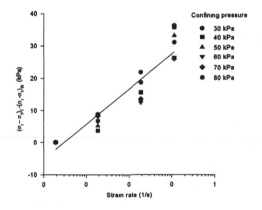

(b)

Figure 3.9 Shear difference at failure versus strain rate.

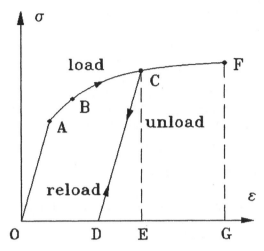

Fig. 3.10 A stress-strain curve of a typical metal curve.

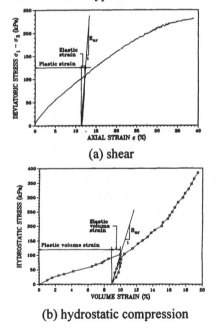

(a) shear

(b) hydrostatic compression

Figure 3.11 Soil stress-strain curves with unloading and reloading. (From *Canadian Agricultural Engineering,* L. Chi et al. 1993. Reproduced with permission of Canadian Society of Agricultural Engineers, Saskatoon, SK, Canada.)

One distinguished feature of soil manipulation in agricultural production is that soil undergoes substantial deformation before the external load is removed, and a large amount of irreversible deformation remains after the removal of load. Fig. 3.11 illustrates a typical loading-reloading behavior of soil after undergoing a noticeable amount of deformation in triaxial shear and hydrostatic compression tests. In both cases, only a small amount of deformation is reversible, i.e., elastic, and most part of deformation is irreversible, i.e., plastic. This indicates that the plastic deformation dominates the soil deformation in agricultural operations and more attention is required for these phenomena.

3.3.1 Classification of Plastic Models

Constitutive models associated to plasticity can generally be divided into two types:
- *Rigid, perfectly plastic models*
- *Elasto-plastic models*

These models will be introduced below in sequence.

Rigid, perfectly plastic models

The simplest form of plasticity is a rigid, perfect plastic model, as illustrated in Fig. 3-12a. In the model, there are no elastic or recovered strains and no changes related to a fixed yield surface. Before stresses reach the yield point, soil is assumed not to deform like a rigid object; beyond the yield point, pure plastic deformation is assumed to occur in soil without any limitation. This model can be used to estimate the limit capacity or load of soil, but generally it oversimplifies the stress-strain behavior of agricultural soils.

A little more realistic model is elastic, a perfect plastic model in which a certain amount of elastic strains is allowed before stresses reach the yield point, as depicted in Fig. 3.12b and marked by 1. Other aspects of the latter is similar to those of the former.

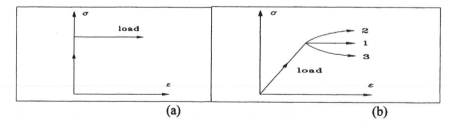

(a)   (b)

Fig. 3.12 Types of stress-strain model. (a) rigid, perfect plastic; (b) elasto-plastic: (1) perfectly plastic, (2) strain hardening, (3) strain softening.

Elasto-Plastic Models
There are some elastic and some plastic strains in the stress-strain relation for agricultural soils. According to different behavior after stresses have reached the yield stress, there exist three cases: perfectly plastic, strain hardening and strain softening, as in Fig. 3.12b and marked by 1, 2 and 3, respectively. This type of model can be used to analyze all cases with or without reloading, at the cost of complexity in constitutive models and FEM calculations.

3.3.2 Underlying Assumptions of Soil Plasticity

Unlike the stress-strain relation of elasticity, the relations arising from plasticity theory usually are incremental, that is, the stress and strain are entirely related by their incremental or differential components. The incremental stress-strain relations for an elasto-plastic material are established on the basis of three basic assumptions: yield and failure criteria, flow rule and hardening law.

3.3.2.1 Yield and Failure Criterion
A *yield criterion f* is a function of stress, strain and other parameters such that when $f < k$ soil is elastic and when $f = k$ soil is in a plastic state. Here, k is a yielding constant depending on material properties. The function f cannot be greater than k; this requirement is known as the *consistency condition*. The yield criterion f designates a *yield surface* in stress space which divides the region into two parts. Inside the yield surface, only elastic strains occur; outside the surface, both elastic and plastic strains are possibly generated.

A failure criterion F is a function of stress, strain and other parameters such that when $F < k_0$ soil failure does not occur and when $F = k_0$ soil is in a failure state where soil plastic deformation becomes unlimited. Here, k_0 is a failure constant depending on material properties. The failure function F corresponds to a failure surface in stress space which is the bound or limit to the yield surface. A yield surface must lie inside or, at most, coincide with the failure surface. The shapes of yield and failure surfaces are usually defined to be similar.

One main difference between metal and soil deformations is that with metals, only shear plastic strain is possible; with soils, both shear and volumetric plastic strains may occur. Therefore, there should be yield criteria taking account of shear and volumetric compression cases.

Shear-Type Yield and Failure Criteria
The existing yield and failure criteria can be classified into two categories: non-frictional and frictional models, according to whether models take account of frictional components to their shear strengths.

Non-frictional models
♦ Tresca's yield criterion
This criterion states that plastic strain occurs when the maximum shear stress reaches a certain value k (soil shear strength), as shown in the following equation:

$$f_1(\sigma_{ij}) = \frac{\text{Max}(|\sigma_1 - \sigma_2|, |\sigma_2 - \sigma_3|, |\sigma_3 - \sigma_1|)}{2} \tag{3.50}$$

This equation represents a prism with a hexagonal cross-section, centered on the hydrostatic axis ($\sigma_1 = \sigma_2 = \sigma_3$) in principal stress space, as shown in Fig. 3.13(a).

♦ von Mises's yield criterion
It states that plastic strain occurs when the maximum strain energy due to shearing equals a critical value k which depends on material properties. This criterion takes account of the contribution of the intermediate principal stress and is more easily handled mathematically, compared to the Tresca's criterion. The detailed algebraic form is:

$$f_1(\sigma_{ij}) = J_2 = \frac{1}{2} s_{ij} s_{ij}$$

$$= \frac{1}{6}[(\sigma_{xx} - \sigma_{yy})^2 + (\sigma_{yy} - \sigma_{zz})^2 + (\sigma_{xx} - \sigma_{zz})^2] + \sigma_{xy}^2 + \sigma_{yz}^2 + \sigma_{xz}^2 \tag{3.51}$$

This equation corresponds to a cylindrical surface centered on the hydrostatic axis in principal stress space, as illustrated in Fig. 3.13(b).

Both Tresca's and von Mises's yield criteria were originally developed for metals. They are not very applicable to soils. It should be noted that the critical value k for a yield criterion should be less than or, at the most, equal to the critical value k_0 for the corresponding failure criterion.

Frictional models
♦ Mohr-Coulomb failure criterion
It is based on the following Mohr-Coulomb law:

$$\tau = c + \sigma \tan \phi \tag{3.52}$$

which can be transformed into the following form in the three-dimensional stress space:

$$F_1(\sigma_y) = \sigma_1 - \sigma_3 - [\sin\phi(\sigma_1 + \sigma_3) + 2c\cos\phi]$$
$$k_0 = 0 \tag{3.53}$$

This equation is equivalent to an irregular hexagonal pyramid surface centered on the hydrostatic axis in principal stress space shown in Fig. 3.13c. One deficiency of the Mohr-Coulomb criterion for three-dimensional analysis is that it has corners in its hexagonal section.

◆ Drucker-Prager's yield criterion
Drucker and Prager (1952) used a conical surface to round off the hexagonal pyramid surface for the mathematical convenience as shown in Fig. 3.13d, and proposed using the following modified form of the Mohr-Coulomb law that takes account of all the principal stresses:

$$f_1(\sigma_{ij}) = \alpha\, I_1 + J_2^{1/2}$$
$$= \alpha\, \sigma_{ii} + \left(\frac{1}{2}s_{ij}s_{ij}\right)^{1/2} \tag{3.54}$$

where:

$$I_1 = \sigma_{xx} + \sigma_{yy} + \sigma_{zz}$$

However, there is evidence that the original Mohr-Coulomb law fits the experimental data better (Bishop, 1966).

◆ Lade yield criterion
Lade (1977) proposed a special yield criterion for cohesionless soils. Its function in principal stress space is:

$$f_1(\sigma_{ij}) = \left(\frac{I_1^3}{I_3} - 27\right)\left(\frac{I_1}{p_a}\right)^m \tag{3.55}$$

where:

$$I_3 = \sigma_{xx}\sigma_{yy}\sigma_{zz} + \tau_{xy}\tau_{yz}\tau_{zx} + \tau_{yx}\tau_{zy}\tau_{xz}$$
$$- (\sigma_{xx}\tau_{yz}\tau_{zy} + \sigma_{yy}\tau_{zx}\tau_{xz} + \sigma_{zz}\tau_{xy}\tau_{yx})$$

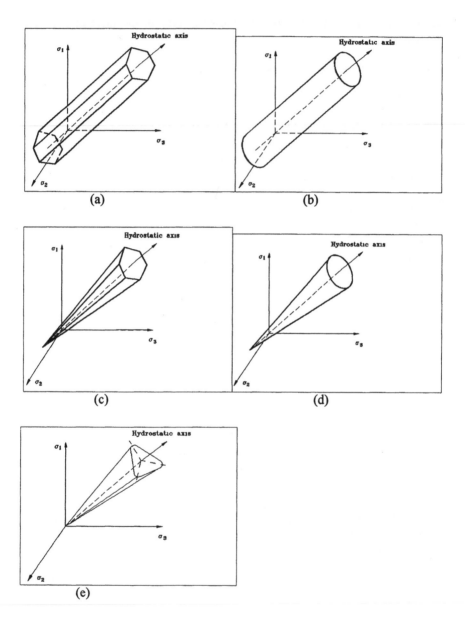

Figure 3.13 Shear-type yield and failure criteria. (a) Tresca's yield surface; (b) von Mises's yield surface; (c) Mohr-Coulomb failure surface; (d) Drucker-Prager's yield surface; (e) Lade yield surface.

Equation (3.55) corresponds to a continuous yield surface in Fig. 3.13e, and is the only one that attempts to incorporate the observed differences in soil behavior along different stress paths, radial from the hydrostatic axis. The price for this is some complexity.

The value of f_1 in equations (3.54) and (3.55) can be used to be compared with the value of k, a constant depending on material properties, to determine whether soil is in the yield state or not. If another critical value k_0 ($k_0 > k$) is used to replace k, the failure criterion is obtained.

Volumetric-Type Yield Criteria
It should be noted that there is no failure criterion for hydrostatic compression, because the yielding in this case does not result in eventual failure. Failure is entirely controlled by the failure surface in shear cases.

◆ Drucker-Gibson-Henkel yield criterion
Drucker et al. (1955) first noted that the closing of the yield surface on the hydrostatic axis was the logic representation of plastic volumetric strain of soils. They proposed a bullet-shaped yield surface consisting of a Mohr-Coulomb surface and a cap which passed through the hydrostatic axis, as in Fig. 3.14a. The cap translates along this axis with its position determined by the preconsolidation stress. However, this surface is not formulated and no careful experimental data were provided to support this criterion.

◆ Cam-Clay yield criterion
The Cam-Clay model (Roscoe et al. 1958) introduced the concept of critical state and the formulation of basic energy dissipation expressions in order to develop an equation for the yield surface. It is written in *p-q* space as:

$$f_2(\sigma_{ij}) = \frac{q}{p \ln(p_c / p)}$$

$$k = M$$

(3.56)

The yield locus represented by this equation is illustrated in Fig. 3.14b. Two dissatisfactions associated with this model are (1) experimental data indicated that the shear strains predicted by the Cam-Clay model were too high at low stress ratios (Britto and Gunn, 1987); (2) Koiter (1953) showed that the plastic strain increment vector at the slope discontinuing point, *A*, must lie within a sector of possible directions, as shown in Fig. 3.14b.

♦ Roscoe and Burland's yield criterion

Roscoe and Burland (1968) corrected the deficiencies of the original version of Cam-Clay yield criterion by introducing the following function in p-q space:

$$f_2(\sigma_{ij}) = \frac{q^2}{(p_c\, p - p^2)}$$
$$k = M^2$$

$$(3.57)$$

This Equation represents an ellipse in p-q space, as shown in Fig. 3.14c. Prevost and Hoeg (1975) used a critical state line in their model, but defined two yield surfaces, one for volumetric and shear deformation and the other for shear deformation alone. The Cam-Clay model has been used in various forms by many investigators such as Adachi and Okano (1974), Pender (1977) and Wilde (1979).

♦ Desai-Siriwardane yield criterion

Desai and Siriwardane (1984) generalized the Cam-Clay model to three-dimensional space by proposing the following function

$$(3.58)$$

♦ = Weidlinger cap model (1971)

This model was developed by DiMaggio and Sandler (1971), Sandler and Baron (1979) and Sandler et al. (1976). The model consists of a stationary failure surface and a strain hardening moving cap that is situated between the failure surface and the hydrostatic axis. The hardening moving cap can be expressed by the following function:

$$f_2(\sigma_{ij}, \varepsilon^p) = (I_1 - L)^2 + R^2 J_2 - (X - L)^2$$
$$k = 0$$

$$(3.59)$$

The yield surface represented by equation (3.59) is illustrated in Fig. 3.14d. The volumetric plastic strain governs the movement of the cap according to a semi-empirical hardening rule.

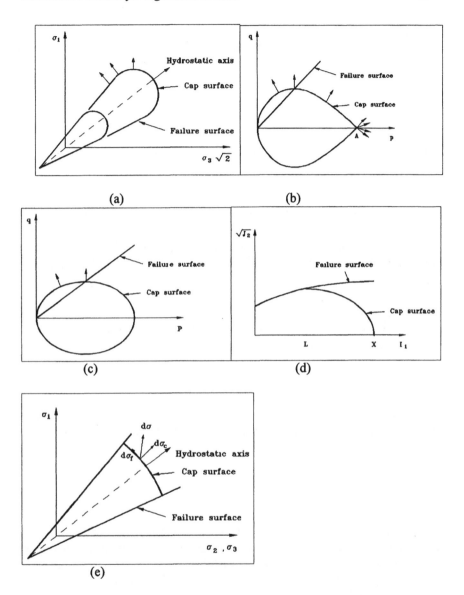

Figure 3.14 Volumetric-type yield criteria. (a) Drucker-Gibson-Henkel yield surface; (b) Cam-Clay yield criterion surface; (c) Roscoe and Burland's yield surface; (d) Sandler-DiMaggio-Baladi yield surface; (e) Lade yield surface.

♦ Lade yield criterion (1977)
The function of this criterion can be written in terms of the first and the second
stress invariant, I_1 and I_2, as follows:

$$f_2(\sigma_{ij}) = I_1^2 + 2I_2 \tag{3.60}$$

where:

$$I_2 = \tau_{xy}\tau_{yx} + \tau_{yz}\tau_{zy} + \tau_{zx} - (\sigma_{xx}\sigma_{yy} + \sigma_{yy}\sigma_{zz} + \sigma_{zz}\sigma_{xx})$$

$k =$ the maximum value of $I_1^2 + 2I_2$ the soil ever experienced

If the value of $I_1^2 + 2I_2$ increase beyond its current value k, the soil work-
hardens and volumetric plastic strain are generated, and then k is updated. The
equation (3.60) corresponds to a sphere surface centered in the origin of the
principal stress space and connected with a shear-type yield surface, as in Fig.
3.14e.

3.3.2.2 Flow Rule

In most models of soil plasticity, it is assumed that the total soil strain increment
is composed of elastic and plastic strain increments:

$$d\varepsilon_{ij} = d\varepsilon_{ij}^e + d\varepsilon_{ij}^p \tag{3.61}$$

The direction of elastic strain increment generally coincides with the direction
of the stress increment, and the magnitude of elastic strain increment is
determined by the following equation:

$$d\varepsilon_{ij}^e = \frac{ds_{ij}}{2G} \tag{3.62}$$

The direction of plastic strain increment is usually not coaxial with the
direction of the stress increment, and its magnitude is not as easily determined
as that of elastic strain increment. Therefore, a *flow rule* is proposed to
determine the direction and relative magnitude of plastic strain increment after
the yield surface is contacted.

Direction of Plastic Strain Increment

Since plastic flow is somehow similar to fluid flow, the use of a *plastic potential
function*, $g(\sigma_{ij})$, is a natural way to describe a vector quantity which depends
only on the location of a point in space. A plastic potential function defines a

plastic potential surface in stress space. The direction of the plastic strain increment is assumed to be the direction of maximum gradient of the plastic potential function at the point where a stress state contacts it. In other words, this direction is at a right angle to the plastic potential surface at the stress point, as shown in Fig. 3.15.

Magnitude of Plastic Strain Increment
According to Drucker (1959), the plastic strain increment is given by a scalar constant β times a vector m_{ij} normal to the potential surface at the stress point:

$$d\varepsilon_{ij}^P = \beta\, m_{ij} = \beta\, \frac{\partial\, g(\sigma_{ij})}{\partial \sigma_{ij}} \tag{3.63}$$

If the plastic potential function is known, the magnitude of plastic strain increment $d\varepsilon_{ij}^P$ is entirely dependent on the constant β which can be determined by the plastic potential function and a work-hardening law. This will be discussed in more detail in the next section.

At present, there exists no definite relationship between the plastic potential function $g(\sigma_{ij})$ and the yield function $f(\sigma_{ij})$. Two possible relations between functions g and f lead to the following classification of the flow rules.

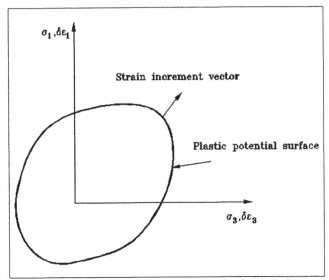

Figure 3.15 Plastic potential surface and plastic strain increment.

◆ Associative flow rule

When $g = f$, the flow rule is called associative (or associated). In this case, the condition of *normality* holds because the plastic strain increment is normal to the yield surface. More usually, the flow rule corresponding to a volumetric-type yield criterion is assumed to be associative such that the uniqueness of the solution to a problem is ensured and the constitutive relationship is simplified.

◆ Non-associative flow rule

When $g \neq f$, the flow rule is called non-associative (or non-associated), and the condition of normality does not hold. One objection to the friction-type failure criteria is that the associative flow rule implies very large extensional volume plastic strains which were not observed experimentally. Therefore, if a friction-type yield criterion is used, a non-associated flow rule should be adopted. For example, Lade (1977) used a plastic potential function with a different form compared to equation (3.60):

$$g_1(\sigma_{ij}) = I_1^3 - (27 + \eta_2(p_a / I_1)^m)I_3 \tag{3.64}$$

3.3.2.3 Yield Surface Hardening and Work-Hardening Law

Yield Surface Hardening

A yield surface is encountered in a work-hardening soil when the first plastic deformation occurs. Then if the stress continues to increase, the yield surface will expand away from the hydrostatic axis until it meets the failure surface. *Yield surface hardening* is mainly concerned with how the yield surface grows, if deformational stress of soil increases. Prager (1959) proposed two possible ways in which the yield surface could develop: isotropic and kinematics hardening.

◆ Isotropic hardening

Isotropic hardening is defined as an equidimensional growth of the yield surface in all directions caused by a stress state encountering it at any position and continuing to increase, as shown in Fig. 3.16a.

This hardening is not consistent with the experimental behavior of soils, but is simple for the mathematical implementation and can be used in the case where the loading applied to soil is monotonic.

◆ Kinematic hardening

Kinematic hardening is defined as a translation of the yield surface with a stress point when this point increases beyond the current yield surface, as in Fig. 3.16b.

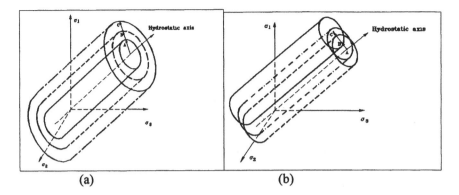

(a)　　　　　　　　　　　　　　(b)

Figure 3.16 Hardening laws. (a) Isotropic; (b) kinematic.

Compared to the isotropic hardening, the kinematic yield surface is a more realistic representation of real metal and soil behavior, but is difficult to implement mathematically because the position of the yield surface must be tracked and remembered.

In addition to the above hardenings, there is a combined isotropic and kinematic hardening which takes account of both translation and expansion of the yield surface. Generally speaking, a hardening law is incorporated into the yield surface function by:

$$f(\sigma_{ij}, h) = 0 \qquad (3.65)$$

where, h is a hardening parameter, a scalar function of the plastic strain which defines the size of the yield surface. For instance, if the von Mises yield surface displaces a certain distance from the hydrostatic axis such that a deviate stress h_{ij} is lying on the new axis of the von Mises cylinder, then equation (3.51) should be rewritten as:

$$f_1(\sigma_{ij}, h_{ij}) = k^2 - \frac{1}{2}(s_{ij} - h_{ij})(s_{ij} - h_{ij}) \qquad (3.66)$$

Work-Hardening Law
Work-hardening law is termed as the relationship between the work used to produce plastic yield (i.e., plastic work W_p) and the stress level. More usually, the work-hardening law of soil is established on the basis of experimental data (Lade, 1975, 1977). The magnitude of plastic strain increments caused by a given stress increment can be calculated using the relation of the work-hardening law and the plastic potential function (Hill, 1950). For example, if

$g(\sigma_{ij}) = I_1^3 - k_2 I_3$, then this plastic potential function is a homogeneous function of degree 3 and β in equation (3.63) can be determined by the following equation:

$$\beta = \frac{dW_p}{3g(\sigma_{ij})} \tag{3.67}$$

3.3.3 Complete Constitutive Model and Incremental Procedure

The different combinations of foregoing yield and failure functions, hardening laws and flow rules can generate a set of various constitutive models. As an example, the complete Lade model is outlined below.

3.3.3.1 Case Model Proposed by Lade (1977)

In this elasto-plastic model, the total strain increments, $d\varepsilon$, are divided into an elastic component, $d\varepsilon^e$, a plastic collapse component associated with a cap-type yield surface, $d\varepsilon^c$, and a plastic expansive component associated with a conical yield surface with the apex at the origin of the stress space, $d\varepsilon^P$, as illustrated by the following equation:

$$d\varepsilon = d\varepsilon^e + d\varepsilon^c + d\varepsilon^P \tag{3.68}$$

where:

$$d\varepsilon^T = \left[d\varepsilon_{xx}, d\varepsilon_{yy}, d\varepsilon_{zz}, 2d\varepsilon_{yz}, 2d\varepsilon_{zx}, 2d\varepsilon_{xy} \right] \tag{3.69}$$

Each of the strain increments in equation (3.68) is calculated separately below.

♦ Elastic strain increments
Hooke's law is used as follows:

$$d\sigma = \mathbf{C}^e \, d\varepsilon^e \tag{3.70}$$

where:

$$d\sigma^T = \left[d\sigma_{xx}, d\sigma_{yy}, d\sigma_{zz}, d\tau_{yz}, d\tau_{zx}, d\tau_{xy} \right] \tag{3.71}$$

For an isotropic material,

$$
\mathbf{C}^e = \frac{E}{(1+v)(1-2v)}
\begin{bmatrix}
1-v & v & v & 0 & 0 & 0 \\
v & 1-v & v & 0 & 0 & 0 \\
v & v & 1-v & 0 & 0 & 0 \\
0 & 0 & 0 & \dfrac{1-2v}{2} & 0 & 0 \\
0 & 0 & 0 & 0 & \dfrac{1-2v}{2} & 0 \\
0 & 0 & 0 & 0 & 0 & \dfrac{1-2v}{2}
\end{bmatrix}
\tag{3.72}
$$

♦ Plastic expansive strains

These strains are determined on the basis of plastic stress-strain theory with a conical yield surface predefined as:

$$
f_p = \left(\frac{I_1^3}{I_3} - 27\right)\left(\frac{I_1}{p_a}\right)^m
\tag{3.73}
$$

$$
f_p = k_1 \quad at\ failure
$$

where: k_1 and m = constants to be determined for specific soils at the desired density

The following form of plastic potential function is chosen:

$$
g_p = I_1^3 - (27 + k_2(p_a / I_1)^m)I_3
\tag{3.74}
$$

where: k_2 = constant for a given value of f_p in equation (3.73) and confining pressure, leading to a non-associative flow rule

The work-hardening or work-softening law is implemented by the following equation:

$$
f_p = a\exp(-bW_p)(W_p / p_a)^{1/q}, \quad q > 0
\tag{3.75}
$$

where: a, b, and q = constants for a given value of the confining pressure

◆   Plastic collapse strains
The collapse strains are determined on the basis of plastic stress-strain theory
with a cap yield surface predefined by:

$$f_c = I_1^2 + 2I_2 \tag{3.76}$$

The corresponding plastic potential function is of identical form to the yield
function:

$$g_c = I_1^2 + 2I_2 \tag{3.77}$$

This results in an associated flow rule, as implied by the normality condition
used for this surface. The work-hardening law used with the spherical yield
surface is chosen as:

$$f_c = p_a^2 \left(\frac{1}{C}\right)^{1/H} \left(\frac{W_c}{p_a}\right)^{1/H} \tag{3.78}$$

where: C and H = constants

3.3.3.2 Incremental Procedure

Like the foregoing case model, many soil plastic constitutive models are often
not of the form ready for the direct implementation of matrix computation for
material non-linearity, yielding and failure in a finite element analysis. Lade and
Nelson (1984) developed a procedure for establishing the incremental stiffness
matrix for an elasto-plastic model with multiple, intersecting yield surfaces. The
flow rule can be either associative or non-associative with respect to each yield
surface. Even though the Lade-Nelson method may take account of up to five
independent yield surfaces, for the simplicity of discussion, the incremental
procedure for an elasto-plastic model with only two yield surfaces is considered
here.

The total strain increment, $d\varepsilon$, is assumed to be the sum of the following
three components:

$$d\varepsilon = d\varepsilon^e + d\varepsilon^{p1} + d\varepsilon^{p2} \tag{3.79}$$

where: $d\varepsilon^e$ = the elastic strain increments determined by equations (3.69) and
(3.70)

de^{pk} = the plastic strain increments associated with the kth surface ($k = 1$ or 2)

de^{pk} in equation (3.79) is determined by a work-hardening plasticity theory which involves:

1. A yield criterion. It may be expressed by a yield function, $f_k(\sigma, \varepsilon^{pk})$, for a work-hardening soil.
2. A flow rule. It assumes that the plastic strain increments and stresses are related by the following specific form:

$$d\varepsilon^{pk} = \beta_k \frac{\partial g_k(\sigma)}{\partial \sigma} \tag{3.80}$$

The derivative of the kth plastic potential can be represented as a vector form:

$$\left(\frac{\partial g_k}{\partial \sigma}\right)^T = \left[\frac{\partial g_k}{\partial \sigma_{xx}} \quad \frac{\partial g_k}{\partial \sigma_{yy}} \quad \frac{\partial g_k}{\partial \sigma_{zz}} \quad 2\frac{\partial g_k}{\partial \tau_{yz}} \quad 2\frac{\partial g_k}{\partial \tau_{zx}} \quad 2\frac{\partial g_k}{\partial \tau_{xy}}\right] \tag{3.81}$$

The basic point in the Lade-Nelson incremental procedure is to treat the increments of all yield function f_k as zero during any increments in stress or strain, if soil is already yielded. Consequently, with regards to each yield function:

$$\left(\frac{\partial f_k}{\partial \sigma}\right)^T d\sigma + \left(\frac{\partial f_k}{\partial \varepsilon^{pk}}\right)^T d\varepsilon^{pk} = 0 \tag{3.82a}$$

where:

$$\left(\frac{\partial f_k}{\partial \sigma}\right)^T = \left[\frac{\partial f_k}{\partial \sigma_{xx}} \quad \frac{\partial f_k}{\partial \sigma_{yy}} \quad \frac{\partial f_k}{\partial \sigma_{zz}} \quad 2\frac{\partial f_k}{\partial \tau_{yz}} \quad 2\frac{\partial f_k}{\partial \tau_{zx}} \quad 2\frac{\partial f_k}{\partial \tau_{xy}}\right] \tag{3.82b}$$

$$\left(\frac{\partial f_k}{\partial \varepsilon^{pk}}\right)^T = \left[\frac{\partial f_k}{\partial \varepsilon_{xx}^{pk}} \quad \frac{\partial f_k}{\partial \varepsilon_{yy}^{pk}} \quad \frac{\partial f_k}{\partial \varepsilon_{zz}^{pk}} \quad \frac{\partial f_k}{\partial \varepsilon_{yz}^{pk}} \quad \frac{\partial f_k}{\partial \varepsilon_{zx}^{pk}} \quad \frac{\partial f_k}{\partial \varepsilon_{xy}^{pk}}\right] \tag{3.82c}$$

Substituting equations (3.70) and (3.79) into equation (3.82a) gives:

$$\left(\frac{\partial f_k}{\partial \sigma}\right)^T \mathbf{C}^e \left(d\varepsilon - d\varepsilon^{p1} - d\varepsilon^{p2}\right) + \left(\frac{\partial f_k}{\partial \varepsilon^{pk}}\right)^T d\varepsilon^{pk} = 0 \quad k = 1,2 \tag{3.83}$$

The substitution of equation (3.80) into equation (3.83) yields:

$$\left(\frac{\partial f_k}{\partial \sigma}\right)^T \mathbf{C}^e \left(d\varepsilon - \beta_1 \frac{\partial g_1}{\partial \sigma} - \beta_2 \frac{\partial g_2}{\partial \sigma}\right) + \beta_k \left(\frac{\partial f_k}{\partial \varepsilon^{pk}}\right)^T \frac{\partial g_k}{\partial \sigma} = 0 \quad k = 1,2 \tag{3.84}$$

Two equations in equation (3.84) are sufficient to solve two unknowns, β_1 and β_2. Equation (3.84) can be rewritten in a matrix form:

$$\left(\frac{\partial f}{\partial \sigma}\right)^T \mathbf{C}^e \left(d\varepsilon - \frac{\partial g}{\partial \sigma}\beta\right) + \mathbf{D}\beta = 0 \tag{3.85}$$

Rearranging equation (3.85) leads to the solution for the unknown column matrix β as follows:

$$\beta = \mathbf{L}^{-1} \left(\frac{\partial f}{\partial \sigma}\right)^T \mathbf{C}^e \, d\varepsilon \tag{3.86}$$

where:

$$\mathbf{L} = \left(\frac{\partial f}{\partial \sigma}\right)^T \mathbf{C}^e \frac{\partial g}{\partial \sigma} - \mathbf{D} \tag{3.87}$$

With the consideration of equations (3.79), (3.80) and (3.86), equation (3.70) can be transformed to:

$$d\sigma = \left(\mathbf{C}^e - \mathbf{C}^e \frac{\partial g}{\partial \sigma} \mathbf{L}^{-1} \left(\frac{\partial f}{\partial \sigma}\right)^T \mathbf{C}^e\right) d\varepsilon \tag{3.88}$$

in which, the first term on the right-hand side of the above equation is the desired elasto-plastic constitutive matrix, i.e.,

$$\mathbf{C}^{ep} = \mathbf{C}^e - \mathbf{C}^e \frac{\partial g}{\partial \sigma} \mathbf{L}^{-1} \left(\frac{\partial f}{\partial \sigma}\right)^T \mathbf{C}^e \tag{3.89}$$

With the availability of the incremental constitutive matrix in equation (3.89), an incremental tangential modulus method in FEM can be used to solve a variety of elasto-plastic problems. The total loading to a system is first divided into many small increments. During each increment, the constitutive matrices as well as element stiffness matrices, are updated according to the status of stress and the position and shape of yield and failure surfaces at the end of the preceding increment.

3.3.4 Case Study: An Elasto-Plastic Model for an Agricultural Cohesive Soil

A stress-strain model was developed based upon the following two objectives (Chi et al. 1993b):
1. A model to account for elasto-plastic behavior of soil, and
2. A model which is convenient to apply in a finite element analysis.

Since considerable plastic strain occurs during triaxial as well as hydrostatic loading, it is essential to formulate an elasto-plastic model to predict stress-strain relationship. Based upon the two objectives established above, the model is developed using the principles of Lade's model (Lade and Duncan, 1975). Since Lade's model was developed for only cohesionless soil, the yield surfaces are modified to account for the effect of soil cohesion.

3.3.4.1 Model Development

Elastic and Plastic Strains
In the development of the model, it is assumed that both elastic and plastic strains occur throughout the loading process from the beginning. The total plastic strain is divided into two parts: (1) the plastic strain associated with a cap-hardening yield surface and (2) the plastic strain associated with a conical yield surface. The total strain increment can be written as follows:

$$d\varepsilon = d\varepsilon^e + d\varepsilon^c + d\varepsilon^f \tag{3.90}$$

where: $d\varepsilon$ and $d\varepsilon^e$ = column matrices of total and elastic strain increments, respectively.

$d\varepsilon^c$ and $d\varepsilon^f$ = plastic strain increment vector associated with cap-type and conical yield surfaces, respectively

The three strain components in equation (3.90) are calculated separately, the elastic strain $d\varepsilon^e$ by Hooke's law; $d\varepsilon^c$ by a plastic stress-strain theory involving a cap-type yield surface; and the $d\varepsilon^f$ by a plastic stress-strain theory involving the cohesional-frictional failure surface.

Yield Functions

Two most commonly used failure criteria for cohesive soils are: (1) the Mohr-Coulomb model, which states that soil fails under a shear stress dependent on cohesion and friction, and (2) the Drucker-Prager model (Drucker and Prager, 1952), which states that soil fails under maximum distortion energy. Since the Mohr-Coulomb model is not a smooth continuous function in three-dimensional cases (Nayak and Zienkiewicz, 1972), it is difficult to employ for constructing a plastic stress-strain relationship. Therefore, the Drucker-Prager model is used as the failure criterion, as given by the following equations:

$$\alpha I_1 + \sqrt{J_2} - k = 0 \tag{3.91}$$

where: I_1 = the first stress invariant
J_2 = the second deivatoric stress invariant
k = a parameter determined from test data
α = a dimensionless parameter determined from test data

The Drucker-Prager criterion provides a smooth conical failure surface in principle stress space as shown in Fig. 3.13d, and is described by the following equations:

$$f_f = \frac{J_2}{(\alpha I_1 - k)^2} \qquad f_f = 1, \quad \text{at failure} \tag{3.92}$$

where: f_f = conical yield function

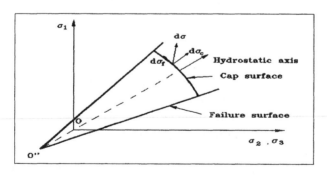

Figure 3.17 Cap yield surface.

Plastic strain is also found during hydrostatic loading in which the conical yield function equation (3.92) is kept constant on the hydrostatic axis of Fig. 3.13d. A cap-type yield criterion is proposed to account for the plastic strain $d\varepsilon^c$ occurring during hydrostatic compression. The cap yield surface is assumed as a sphere with the center at the tip of the conical failure surface (O'' in Fig. 3.17) and thus the cap yield surface is orthogonal to the conical yielding surfaces. Therefore, an increase in stress state can be considered as a combination of the stress changes due to an increase in the conical yield function and in the cap yield function. The cap-type yield function is given by the following equation:

$$f_c = I_1^2 + 2I_2 + \frac{2k}{3\alpha} I_1 \qquad (3.93)$$

where: f_C = cap yield function
I_2 = the second stress invariant

Elastic Constitutive Matrix
The Elastic strain in equation (3.61) is calculated by using Hooke's law. For three dimensional stresses, the elastic constitutive matrix is given by the following Equations:

$$\mathbf{C}^e = \frac{E_{ur}}{(1+v)(1-2v)}
\begin{bmatrix}
1-v & v & v & 0 & 0 & 0 \\
v & 1-v & v & 0 & 0 & 0 \\
v & v & 1-v & 0 & 0 & 0 \\
0 & 0 & 0 & \dfrac{1-2v}{2} & 0 & 0 \\
0 & 0 & 0 & 0 & \dfrac{1-2v}{2} & 0 \\
0 & 0 & 0 & 0 & 0 & \dfrac{1-2v}{2}
\end{bmatrix} \qquad (3.94)$$

where: E_{ur} = elastic modulus
v = Poisson's ratio
$\mathbf{C}^e$ = elasticity matrix

Janbu (1963) proposed a model to describe the modulus as a function of confining pressure by using:

$$E_{ur} = K_{ur} P_a \left(\frac{\sigma_3}{P_a} \right)^n \qquad (3.95)$$

where: E_{ur} = unloading and reloading elastic modulus
         σ_3 = minor principal stress
         p_a = atmospheric pressure
K_{ur} and n = dimensionless parameters determined from test data

Poisson's ratio is determined using the following equation (Zhang et al., 1986):

$$\nu = \frac{3B_{ur} - E_{ur}}{6B_{ur}} \tag{3.96}$$

where: B_{ur} = unloading and reloading bulk modulus

Plastic Strain Associated with the Cap-Type Yield Surface
By using the associated flow rule, the plastic strain increment associated with the cap yield surface is given by the following equation:

$$d\varepsilon^c = \lambda_c \frac{\partial f_c}{\partial \sigma} \tag{3.97}$$

where: λ_c = a positive scalar factor

The work hardening law is applied to determine the scalar λ_c in equation (3.97). Lade (1977) developed a power function to derive a working equation for calculating the total plastic work required to produce collapse strains. However, it was found that the Lade's equation did not fit the data obtained from hydrostatic loading test of cohesive soil, especially when the volumetric strain was greater than 10 percent. An exponential equation proposed by Bailey et al. (1984) for soil compaction showed good accuracy over a wide range of volumetric strain. Therefore, an exponential function is proposed to describe the relationship between the total plastic work required to produce the plastic strain $d\varepsilon^c$ and the degree of hardening expressed by f_c, as given by the following equation:

$$W_c = C\sqrt{f_c}\left[1 - \exp(-p f_c / p_a^2) \right. \tag{3.98}$$

where: W_c = plastic work associated with cap yield surface
              C and p = dimensionless parameters determined from test data

Plastic Strain Associated with the Conical Yield Surface

By using the associated flow rule, the plastic strain associated with the conical yield surface is given by the following equation:

$$d\varepsilon^f = \beta_f \frac{\partial f_f}{\partial \sigma} \tag{3.99}$$

where: β_f = a positive scalar factor

The work hardening law is applied to determine the scalar λ_f in equation (3.99). Lade's model including a work hardening law with the conical yield surface could be used to simulate both work hardening and work softening behavior of materials. Lade's work hardening and softening model can be expressed by:

$$f_f = F_f(W_f) = a e^{-bW_f} \left(\frac{W_f}{P_a} \right)^{\frac{1}{q}} \tag{3.100a}$$

$$a = \left(\frac{e P_a}{W_{fpeak}} \right)^{\frac{1}{q}} \tag{3.100b}$$

$$b = \frac{1}{q W_{fpeak}} \tag{3.100c}$$

$$W_{fpeak} = \theta \, p_a \left(\frac{\sigma_3}{P_a} \right)^{\Lambda} \tag{3.100d}$$

$$q = \gamma + \beta \frac{\sigma_3}{P_a} \tag{3.100e}$$

where: W_f = plastic work associated with conical yield surface done per unit volume
W_{fpeak} = plastic work at f_f = 1
a, b and q = intermediate variable
γ, β, θ and Λ = dimensionless work hardening parameters

Complete Elasto-Plastic Constitutive Matrix

The complete elasto-plastic constitutive relationship includes an elastic component and two work hardening plastic components. By using the Lade-

Nelson method introduced in Section 3.3.3, the complete elasto-plastic model is derived as:

$$d\sigma = \mathbf{C}^{ep} d\varepsilon \tag{3.101a}$$

$$\mathbf{C}^{ep} = \mathbf{C}^e - \mathbf{C}^e \frac{\partial F}{\partial \sigma} \mathbf{L}^{-1} \left(\frac{\partial F}{\partial \sigma} \right)^T \mathbf{C}^e \tag{3.101b}$$

$$\mathbf{L} = \left(\frac{\partial F}{\partial \sigma} \right)^T \mathbf{C}^e \frac{\partial F}{\partial \sigma} + \mathbf{D} \tag{3.101c}$$

$$\frac{\partial F}{\partial \sigma} = \left[\frac{\partial f_c}{\partial \sigma}, \frac{\partial f_f}{\partial \sigma} \right] \tag{3.101d}$$

$$\mathbf{D} = \begin{bmatrix} -\dfrac{\partial f_c}{\partial W_c} \sigma^T \dfrac{\partial f_c}{\partial \sigma} & 0 \\[4mm] 0 & -\dfrac{\partial f_f}{\partial W_f} \sigma^T \dfrac{\partial f_f}{\partial \sigma} \end{bmatrix} \tag{3.101e}$$

where, $\mathbf{C}^e, f_c(\sigma), W_c(f_f), f_f(\sigma)$ and $f_f(W_f)$ are determined by equations (3.94), (3.93), (3.98), (3.92a) and (3.100a), respectively. Therefore, the elasto-plastic constitutive matrix $\mathbf{C}^{ep}$ can be determined if a stress state is known.

3.3.4.2 Test Procedure

In order to investigate the mechanical behavior of agricultural soils, axial compression tests are conducted using a conventional triaxial test apparatus. The confining pressure is applied by pressurized water. The volume change in soil samples is measured by a volumeter developed by Chi and Kushwaha (1989). Axial load, axial displacement, confining pressure and change in volume are measured during tests. The data are sampled by a 21x Campbell Scientific Micro-Logger and then transferred to a computer for analysis.

The test soil is a clay loam. Remolded soil samples 50 mm in diameter and 100 mm in length were used in the tests. A split compaction mold is used to prepare soil specimens and all specimens are compacted in 8 layers at a density of 1434 kg/m^3 and a moisture content of 14.8 %.

An unloading and reloading procedure is applied during the triaxial tests (Fig. 3.18). The curve indicates that soil undergoes a considerable amount of plastic strain during the loading. The elastic modulus of unloading and reloading E_{ur} is used in the elastic constitutive relationship. Three series of tests are conducted in calibrating the model. The first series of triaxial tests is carried out with 5 different axial strain rates of 0.00014, 0.007, 0.036, 0.18 and 0.7/min under a

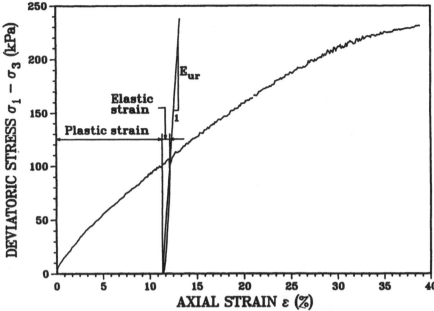

Figure 3.18 Soil stress-strain curve with unloading and reloading. (Chi et al., 1993b.)

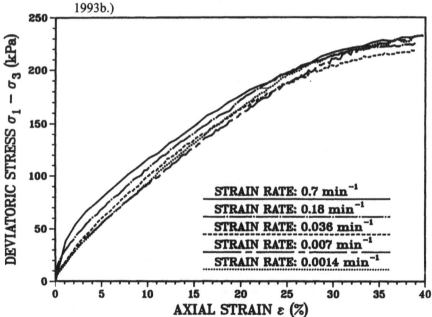

Figure 3.19   Stress-strain curves under different loading speeds with a confining pressure 50 kPa. (Chi et al., 1993b.)

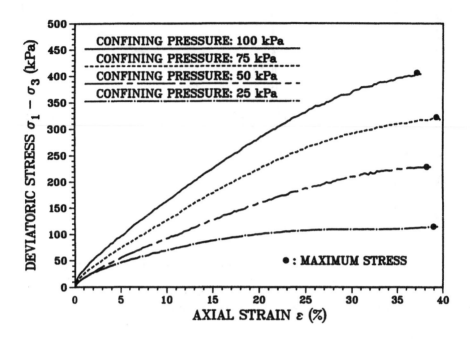

Figure 3.20   Stress-strain curves under different confining pressures. (Chi et al., 1993b.)

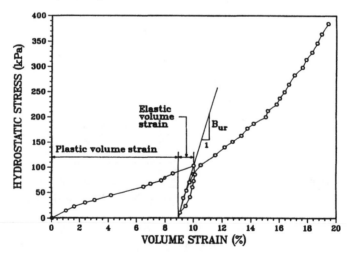

Figure 3.21   Stress-strain curve under hydrostatic compression. (Chi et al., 1993b.)

confining pressure of 50 kPa. The results from this series of tests (Fig. 3.19) shows that the stress-strain relationship is sensitive to loading speeds greater than 0.036/min. The second series of triaxial tests is conducted under four confining pressures of 25, 50, 75 and 100 kPa with a strain rate of 0.007/min (Fig. 3.20). In both the first and second series of tests, the unloading and reloading procedure is applied. However, the unloading and reloading curves are not plotted in Figs. 3.19 and 3.20 for clarity. The final series of tests is conducted under hydrostatic loading ($\sigma_1 = \sigma_2 = \sigma_3$). The results (Fig. 3.21) indicates that plastic strain occurs under hydrostatic loadings. B_{ur} in Fig. 3.21 indicates a bulk elastic modulus of unloading and reloading. All tests are repeated three times.

3.3.4.3 Results

Soil Parameters
The complete elasto-plastic model includes 11 soil parameters: three elastic parameters (K_{ur}, n and ν), two soil failure parameters (α and k), two cap-type working hardening parameters (C and p) and four conical work hardening parameters (γ, β, θ and Λ). All these parameters are determined by using the test data from conventional triaxial compression and hydrostatic compression tests.

According to equation (3.95), a linear regression method is used to estimate two elastic parameters, K_{ur} and n, as shown in Fig. 3.22. K_{ur} and n are calculated to be 284.415 and 0.907, respectively.

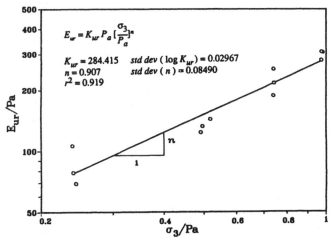

Figure 3.22   Determination of parameters for the elastic behavior. (Chi et al., 1993b.)

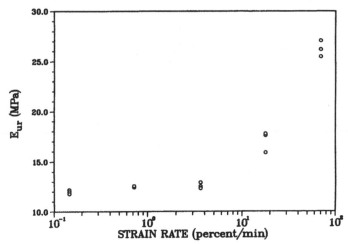

Figure 3.23   Elastic modulus of soil at various loading speeds. (Chi et al., 1993b.)

The elastic modulus, E_{ur}, is affected by strain rate for values greater than 0.036/min, as shown in Fig. 3.23. More tests have to be conducted in order to develop an effective model to include the dynamic effect of strain rate.

The value of bulk modulus, B_{ur}, in equation (3.96) is calculated from test data in hydrostatic loading. The third elastic parameter, Poisson's ratio ν, is calculated as 0.345 by using equation (3.22).

The failure function parameters, α and k, in equation (3.91) are used in both cap and conical yield functions (equations (3.92) and (3.93)). These two parameters are estimated from the maximum value of deviatoric stress (Fig. 3.20). Fig. 3.24 shows the results obtained from the equation (3.91) and the data obtained from the test. Values of α and k are calculated to be 0.334 and 4.242 kPa, respectively.

Fig. 3.25 shows the data obtained from hydrostatic loading tests along with equation (3.98). Two cap-type working hardening parameters, C and p, in equation (3.98) are estimated by using a non-linear regression method. The plastic work is calculated as the difference of total deformation work from the elastic work. C and p are calculated to be 0.0274 and 0.8883, respectively.

Equations (3.100a) through (3.100e) are used to evaluate four conical work hardening parameters (γ, β, θ and Λ). The plastic work associated with the conical yield surface is calculated as the difference of total deformation work from the elastic work and plastic work associated with the cap yield surface. Fig. 3.26 shows the variation of plastic work W_f with the conical yield function ff under different σ_3 confining pressures. Values of γ, β, θ and Λ are calculated to be 1.933, -0.8251, 0.9854 and 0.8492, respectively.

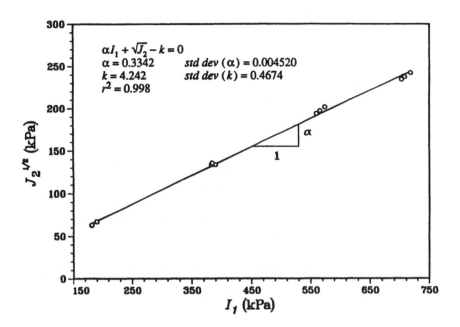

Figure 3.24 Determination of soil failure parameters. (Chi et al., 1993b.)

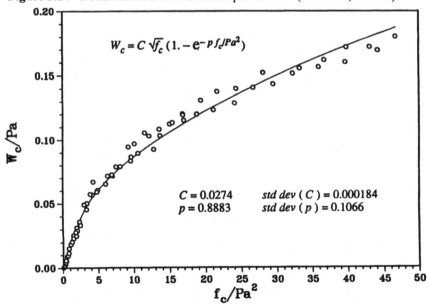

Figure 3.25 Cap plastic work hardening equation. (Chi et al., 1993b.)

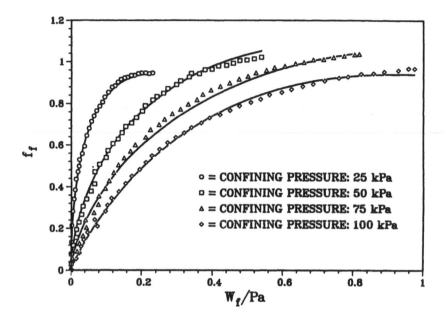

Figure 3.26 Conical plastic work hardening equation. (Chi et al., 1993b.)

Table 3.1 Summary of soil parameters. (Chi et al., 1993b.)

	Parameters	Values	Std. Dev.
Elastic	K_{ur}	284.4 (=$10^{2.454}$)	0.0297
	n	0.907	0.0849
	ν	0.345	
Failure function	α	0.3342	0.00452
	k (kPa)	4.242	0.464
Plastic (cap)	C	0.0274	0.000184
	p	0.8883	0.10664
Plastic (conical)	γ	1.9330	0.0552
	β	-0.8251	0.0821
	θ	0.9854 (=$10^{-0.0064}$)	0.0171
	Λ	0.8492	0.0334

Prediction of the Stress-Strain Relationship

To demonstrate the curve-fitting accuracy of the proposed elasto-plastic model in predicting the deformation behavior of the test soil, calculated values are compared to the data obtained from triaxial tests. Equations (3.90) and (3.101) along with the parameters listed in Table 3.1 are used to calculate theoretical strains. Fig. 3.27 presents the curve-fitting results of the theoretical and experimental stress-strain relationships, expressed by a solid line and four different symbols respectively, under four confining pressures. Fig. 3.28 shows the prediction of stress-strain curve under the hydrostatic compression. The results show a very good curve-fitting accuracy for both axial and hydrostatic compression.

3.3.4.4 Summary and Conclusions

An elasto-plastic stress-strain model for cohesive soil has been developed on the basis of the principles of the Lade model. The Drucker-Prager failure criterion is applied in the model to account for the effect of cohesion. The model includes two yield surfaces: a cap yield surface to calculate the plastic strain caused by the hydrostatic compression and a conical yield function to calculate the plastic strain associated with the Drucker-Prager yield function. In the model, a new work hardening function is proposed for the cap yield surface, which is more suitable for cohesive soils.

The developed model shows a good curve-fitting accuracy for both axial and hydrostatic compression. All parameters can be determined from conventional triaxial tests. The model provides a single form of constitutive relationship during the whole loading period. Therefore, the model can be more easily implemented in finite element analyses compared with the previous critical state and cap models.

3.4 DYNAMIC MODELS

The soil-tool interactions in soil-machine systems are generally dynamic processes, that is, an interaction between a tool and a specific portion of soil is completed within a limited stretch of time.

With a dynamic process, two possible effects might need to be considered in an analysis:

(a) Inertia-effect

Inertia-effect refers to the influence of inertia forces of soil mass on the soil-tool interaction. This effect is considered using a dynamic analysis which will be introduced in Section 5.4.

(b) Rate-effect

Rate-effect refers to the effect of strain rate on the soil-tool interaction. If the soil shows some rheological properties, i.e., its stress-strain relation is time dependent, time should be incorporated into a static constitutive model to form a dynamic model. Below four types of dynamic models will be introduced: viscoelastic, elasto-viscoplastic, rate process, and rate-dependent elastic or elasto-plastic models.

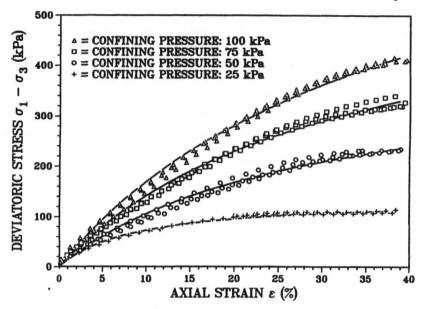

Figure 3.27   Comparison of predicted and measured stress-strain behavior of soil for axial compression. (Chi et al., 1993b.)

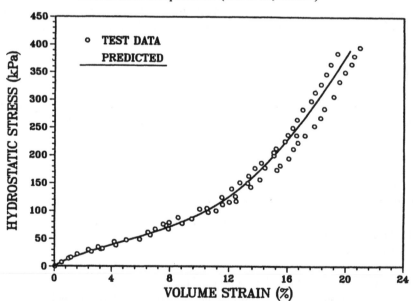

Figure 3.28   Compression of predicted and measured stress-strain behavior of soil for hydrostatic compression. (Chi et al., 1993b.)

3.4.1 Viscoelastic Model

Viscoelasticity denotes the elasticity coupled with viscosity. A material is described as viscous if its strain rate is dependent upon its stress level. Three common types of viscosity have been used in the past:

(a) Newtonian fluid

The Newtonian fluid viscosity denotes that the strain rate is directly proportional to the stress level, i.e.,

$$\frac{d\varepsilon}{dt} = \frac{\sigma}{coefficient\ of\ viscosity} \tag{3.102}$$

where: t = time

When the stress level is low, equation (3.102) can be used to describe the behavior of viscosity. This equation is applicable only to the cases at small stress levels.

(b) Power law

The power law viscosity is one kind of representation of non-linear viscosity as illustrated by the equation:

$$\frac{d\varepsilon}{dt} = \frac{\sigma^k}{\eta} \tag{3.103}$$

where: k, η = material constants. This equation can be used at both high and low stress levels

(c) Exponential

Another type of non-linear viscosity is represented by the exponential function of stress, i.e.,

$$\frac{d\varepsilon}{dt} = \frac{\exp(\sigma)}{\eta} \tag{3.104}$$

This function is likely to adequately represent the viscous behavior over a wide stress range, and is an approximation of the following hyperbolic sine function predicted using a branch of physical chemistry known as rate process theory:

$$\frac{d\varepsilon}{dt} = \frac{\sinh(a\sigma)}{\eta} \qquad\qquad (3.105)$$

Analysis by Mitchell (1976) indicates that $a\sigma$ in the above equation is usually greater than unity, therefore the hyperbolic sine function can be approximately transformed to an exponetinal function as follows:

$$\sinh(a\sigma) \cong \frac{\exp(\sigma)}{2} \qquad\qquad (3.106)$$

A mechanical analogue method was used in the past in developing a variety of viscoelastic models. Elastic and viscous behavior is represented, respectively, using a spring and a dashpot component as in Fig. 3.29a and 3.29b. Another basic component is a slide, as shown in Fig. 3.29c, which represents plastic behavior and will be used in Section 3.4.2. In the case of non-linear viscosity an adjustable dashpot should be used, as in Fig. 3.29d. Three typical viscoelastic models are introduced below.

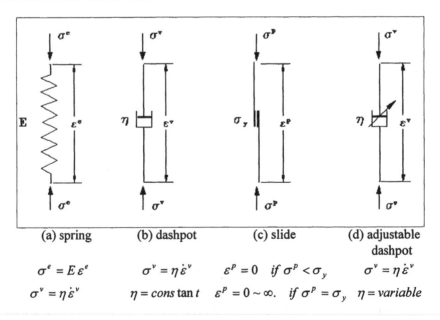

| (a) spring | (b) dashpot | (c) slide | (d) adjustable dashpot |

$$\sigma^e = E\varepsilon^e \qquad \sigma^v = \eta\dot{\varepsilon}^v \qquad \varepsilon^p = 0 \quad if\ \sigma^p < \sigma_y \qquad \sigma^v = \eta\dot{\varepsilon}^v$$

$$\sigma^v = \eta\dot{\varepsilon}^v \qquad \eta = cons\tan t \qquad \varepsilon^p = 0 \sim \infty. \quad if\ \sigma^p = \sigma_y \quad \eta = variable$$

Figure 3.29 Three basic rheological components and one adjustable component.

Kelvin-Voigt Model Fig. 3.30a shows the configuration of this model which is a parallel arrangement of one spring and one dashpot. If the viscosity is considered linear, its stress-strain relation in uniaxial cases is of the form:

$$\sigma = E\varepsilon + \eta\dot{\varepsilon} \qquad (3.107)$$

or

$$\dot{\varepsilon} = \frac{d\varepsilon}{dt} = \frac{1}{\eta}\sigma - \frac{E}{\eta}\varepsilon \qquad (3.108)$$

Maxwell Model It is another simplest viscoelastic model which is a series arrangement of one spring and one dashpot as in Fig. 3.30b. In uniaxial cases, the stress-strain relation for this model with a linear viscous element can be represented by the following equations:

$$\varepsilon = \varepsilon^e + \varepsilon^v \qquad (3.109a)$$

$$\varepsilon^e = \frac{\sigma}{E} \qquad (3.109b)$$

$$\dot{\varepsilon}^v = \frac{\sigma}{\eta} \qquad (3.109c)$$

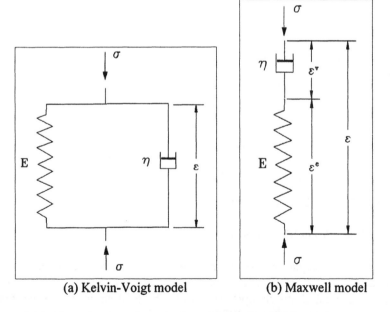

(a) Kelvin-Voigt model (b) Maxwell model

Figure 3.30 Two simplest viscoelastic conceptual models.

Berger's Model It is a more complex combination between elastic springs and viscous dashpots, as illustrated in Fig. 3.31. A series of experiments by Pan et al. (1990) shows this model may give a realistic representation of the rheological behavior of agricultural soils in paddy fields. In uniaxial cases, its stress-strain relation accounting for time effect is given by the following equations:

$$\varepsilon = \varepsilon^e + \varepsilon^{ev} + \varepsilon^v \tag{3.110a}$$

$$\varepsilon^e = \frac{\sigma}{E_M} \tag{3.110b}$$

$$\dot{\varepsilon}^v = \frac{\sigma}{\eta_M} \tag{3.110c}$$

$$\dot{\varepsilon}^{ev} = \frac{d\varepsilon^{ev}}{dt} = \frac{1}{\eta_K}\sigma - \frac{E_K}{\eta_K}\varepsilon^{ev} \tag{3.110d}$$

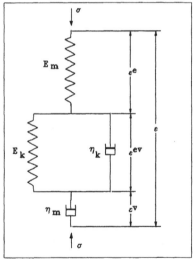

Figure 3.31 The Berger's model.

3.4.2 Elasto-Viscoplastic Model

In an elasto-viscoplastic model, the plastic portion of the model is rate-dependent and the elastic portion is rate-independent. Fig. 3.32 illustrates a simplest uniaxial elasto-viscoplastic model. The total strain can be written as (Zienkiewicz and Humpheson, 1977):

$$\varepsilon = \varepsilon^e + \varepsilon^{vp}$$ (3.111a)

where the elastic strain, ε^e, is given by:

$$\varepsilon^e = E\sigma$$ (3.111b)

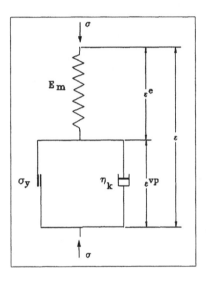

Figure 3.32 A simple uniaxial elasto-viscoplastic model.

and the rate of viscoplastic strain, $\dot{\varepsilon}^{vp}$, is determined by:

$$\frac{\partial}{\partial t}\varepsilon^p = \dot{\varepsilon}^p = a\,\beta(\sigma - \sigma_y)$$ (3.111c)

where: β = fluidity parameter which may be dependent on time and strain
  σ_y = yield stress
  a is determined by:

$$a = \begin{cases} 0 & |\sigma| < |\sigma_y| \\ 1 & |\sigma| \geq |\sigma_y| \end{cases}$$ (3.111d)

These kinds of models can be extended to describe any complex behavior by (a) introducing non-linear dashpots with some arbitrary function to specify the viscoplastic rate,

$$\dot{\varepsilon}^{vp} = \psi(\sigma - \sigma_y)$$

(3.111e)

and (b) introducing a function of yield stress dependent on strain and strain rate, $\sigma_y = \sigma_y(\varepsilon, \dot{\varepsilon})$, and finally (c) placing a number of elastic, plastic and viscous components in series or parallel.

To generalize the uniaxial to multi-dimensional stress-strain behavior, a yield function and a plastic potential function are needed. For instance, equations (3.111a) through (3.111d) are changed to equations (3.112a) through (3.112d) as follows:

$$\varepsilon = \varepsilon^e + \varepsilon^{vp}$$

(3.112a)

$$\varepsilon^e = \left(\mathbf{C}^e\right)^{-1}\sigma$$

(3.112b)

$$\frac{\partial}{\partial t}\varepsilon_{ij}^p = \dot{\varepsilon}_{ij}^p = a\,\beta\frac{\partial g(\sigma_{ij})}{\partial\sigma_{ij}}$$

(3.112c)

$$a = \begin{cases} 0 & f(\sigma_{ij}) < k \\ 1 & f(\sigma_{ij}) = k \end{cases}$$

(3.112d)

where: ε = strain column matrix = $\begin{bmatrix} \varepsilon_{xx} & \varepsilon_{yy} & \varepsilon_{zz} & \varepsilon_{xy} & \varepsilon_{yz} & \varepsilon_{xz} \end{bmatrix}^T$

$\left(\mathbf{C}^e\right)^{-1}$ = inverse matrix of the elastic constitutive matrix

σ = stress column matrix = $\begin{bmatrix} \sigma_{xx} & \sigma_{yy} & \sigma_{zz} & \sigma_{xy} & \sigma_{yz} & \sigma_{xz} \end{bmatrix}^T$

$f(\sigma_{ij})$ = yield function

$g(\sigma_{ij})$ = plastic potential function

Even though the viscoelastic models in Section 3.4.1 and elasto-visco-plastic models in this subsection did serve their purpose to a certain extent, some investigators argued that the choice of the number and combination of basic components (spring, dashpot and slide) was somewhat arbitrary; reasonable agreement between the prediction by some models and actual soil behavior may not necessarily be considered as support for these models, since equations of these models depended only on the mathematical consequences of a particular arrangement of model elements that had been chosen (Singh and Mitchell,

1968). The next subsection will introduce an alternative way to the dynamic model of a soil on a relative sound basis of physics.

3.4.3 Rate Process Model

Agricultural soil is a non-uniform medium in which the size and configuration of solid particles, the granular structure formed by a considerable number of particles, and the interaction among air, water and particles vary tremendously from place to place inside soil. After soil is subject to an external load, the breakage of particles, the air and water squeezed from soil and the destruction of original granular structure due to particles' sliding and rolling over each other add additional complexities. Currently, the development of a model that can account for the foregoing features of a soil appears to be extremely difficult. One reasonable approach to the mechanical behavior of soil is to neglect the subtle changes at different specific places inside soil and to discuss the average behavior of soil on a probability basis.

Statistical mechanics is based on probability and is universally suited in describing various types of mechanical systems. Based upon it, a rate process theory (RPT) was developed by Glasstone et al. (1941) and has been used to study the rate-dependent deformation of a variety of materials such as ceramics (Gibbs and Eyring, 1949), metals (Finnie and Heller, 1959), plastics and rubber (Ree and Eyring, 1958), textiles (Eyring and Halsey, 1948), asphalt (Herrin and Jones, 1963) and concrete (Polivka and Best, 1960). RPT has also been adopted for the study of viscosity, plasticity, friction, lubrication and diffusion (Eyring and Powell, 1944).

As to engineering soils, it has been applied to the prediction of soil behavior by Abdel-Hady and Herrin (1966), Christensen and Wu (1964), Mitchell (1976), Mitchell et al. (1968), Murayama and Shibata (1958), Noble and Demirel (1969) and Keedwell (1984). Even though the basic frame in these previous studies was kept the same, some detailed assumptions and definitions varied among different authors and there existed several ways in abstracting a soil flow unit. In the following, only the assumptions or definitions made by the authors will be introduced. For other approaches, see Mitchell (1976) and Keedwell (1984).

3.4.3.1 Element of Soil Structure

According to the theory of double electric layer formulated by G. Gouy in 1910 and developed by D. L. Chapman in 1913, a solid particle, especially a colloidal particle, is in conjunction with a layer of firmly-bound (absorbed) water comprising positive ions (cations) of water with the thickness of a molecule or a few molecules, as shown in Fig. 3.21 (Vyalov, 1986). Beyond this layer, there exists a layer of loosely-bound (lysisorbed) water which also consists of positive ions of water and flows along the surface of a soil particle in a course of deformation. The firmly-bound and loosely-bound water form a double layer, several hundred angstrom thick (one angstrom = 10^{-10} m). Outside this double layer, free water or air is located. In this study, a soil particle and its surrounding double layer of water are defined as an element of soil structure.

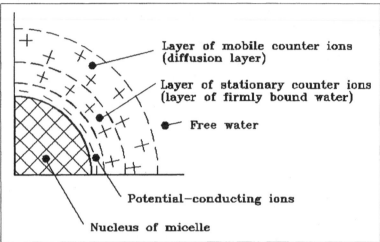

Figure 3.33 Electromolecular forces of a mineral particle and the hydrated envelope (Vyalov, 1986).

3.4.3.2 Force and Potential Energy in Particle Interaction

Each particle is subject to external and internal forces as well as to fields of energy induced by these forces. Internal fields are produced by interparticle forces between soil elements. In general, inter and intra-particle forces in soil include: chemical forces, molecular forces, ionic-electrostatic forces, capillary-electrostatic (Coulomb) forces and magnetic forces (Vyalov, 1986).

The above forces produce fields of energy and form bonds between particles of a dispersed system. For soil with a stable structure, forces of particle interaction can be divided into two types: one is referred to as forces of attraction, f_a, which make particles gather together, and the other is referred to as forces of repulsion, f_r, which prevent the annexation of particles. The forces of attraction and repulsion differ among soil types. To bond particles into a soil structure, the resultant force f and potential energy U of particle interaction in different soil types have some common features, as shown in Fig. 3.34a. When r is large, the absolute value of f_a increases rapidly with the decrease in r while the increase in f_r is relatively small. Consequently, the resultant force $f(r)$ is less than zero and attractive, which makes particles gather together. When r is relatively small, f_r increases more rapidly than f_a with a decrease in r. As a result, $f(r)$ is larger than zero and repulsive, which avoids the annexation of particles. When $r = r_0$, f_a is equal to f_r and f equals zero (Fig. 3.22a). Since $dU(r)/dr = -f(r)$, curve $U(r)$ can be obtained from $f(r)$ (Fig. 3.34b). When $r = r_0$, $dU(r)/dr$ equals zero, and potential energy reaches its minimum value, U_0. Thus, two particles separated by a distance r_0 are in equilibrium.

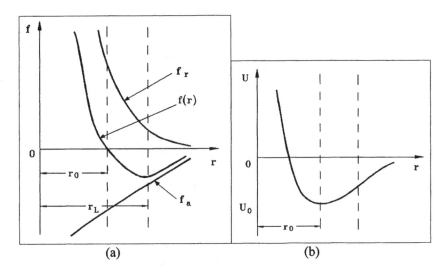

Figure 3.34 Balance between interparticle forces.

3.4.3.3 Idealization of Soil Shear Process

Keedwell (1984) classified the contact zone of soil particles into three types: (1) mineral-to-mineral; (2) mineral-to-mineral and adsorbed water layer, and/or hybrid; (3) adsorbed water layer. The first type is associated with cohesiveless soils, while the rest are related to cohesive soils. Since the rate-effect is more obvious with cohesive soils and generated by viscous fluid elements between soil particles, type 3 will be used as the representative form of the contact zone of soil particles.

As mentioned before, there is usually a layer of firmly-bound water surrounding soil particles which is one or a few molecules in thickness. It is almost impossible for an external force to pull it off the surface of particles, because forces of attraction between a soil particle and its firmly-bound water may reach 1 GPa (Vyalov, 1986). The layer of loosely-bound water is removable from the surface of a particle. During a shear process, a certain amount of normal stress is applied on the double layer of a particle and may force the loosely-bound water off the surface of a particle.

For the sake of simplicity, the boundary interface of a particle in shear process is idealized as a single layer of positive ions of water attached to the particle. The distance between two boundary interfaces, belonging to two adjacent particles respectively, is defined as the distance between these two particles, as shown in Fig. 3.35 This distance decreases with the increase in compressive stresses applied on the double layer of two adjacent particles and cannot become zero due to the repulsion between positive ions in two boundary interfaces. In the equilibrium of external compressive forces, the distance between two adjacent particles equals a quantity similar to r_0 in Fig. 3.34.

The relative displacement between two particles in a shear process can then be simplified as the relative movement of two layers of cations, belonging to two adjacent particles respectively, at a uniform shear plane. The direction and area of this plane depend on contact types between the particles (edge-to-edge, edge-to-face or face-to-face) and the relative position of the particles in rolling over each other in a shear process. Since the contact type and relative position between particles are generally random in soil, actual shear planes between particles in soil may exist in every direction with different contact areas. However, in a statistical sense, an average shear plane between two groups of particles should be definite in both direction and area. For example, the tangential direction of the average shear plane for a specimen in a direct shear test is normal to the axis of the cylindrical specimen and the area of this average shear plane equals the cross sectional area of the soil specimen. If probabilities of the relative movement between two particles in both forward and backward directions along the average shear plane are known, the average shear strain of soil can be calculated on a probability basis. Herein, a uniform shear plane was adopted to simplify the boundary surface formed by two particles, as shown in Fig. 3.35. Cations A and B in Fig. 3.35 represent positive ions of firmly-bound water of group I and group II, respectively.

If external forces are large enough, cation A and group I will move a distance a relative to group II which equals interval distance of two adjacent cations of bound water. In this route, when the line connecting cations A and B is perpendicular to the moving direction, the distance between A and B reaches its minimum value and the force of repulsion becomes maximum. The difference between the potential energy corresponding to maximum repulsion and that in the original equilibrium is called barrier of potential energy. Cations are constrained from movement relative to each other by virtue of energy barriers separating adjacent equilibrium positions by the distance between atoms, as depicted schematically in Fig. 3.36. The displacement of cations to new positions requires sufficient magnitude of activation energy ΔF_a to overcome the barriers. After moving to equilibrium position *2* from position *1*, if external potential is large enough, a cation will continuously move to the position *3, 4* and so on. This forms irreversible macro-deformation of soil.

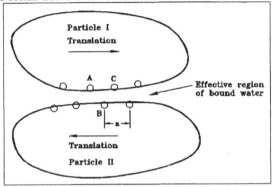

Figure 3.35 Model for the interaction between particles.

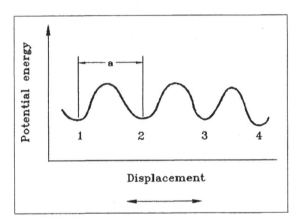

Figure 3.36 Potential barriers and displacement.

3.4.3.4 Application of Statistical Mechanics in Analyzing a Shear Process

Soil consists of a great number of particles per unit volume. Different shapes and sizes of particles and different contact types and relative positions between particles lead to non-uniformly distributed heat energy between different particles in soil. Under such circumstances, the pattern of average distribution of the energy between particles may be described in terms of the Boltzmann distribution law in statistical mechanics (Vyalov, 1986).

- Average translation speed of atoms

A thermo-vibration process takes place both in the lattice of mineral particles and in the molecular structure of firmly-bound water film interlinking particles (Vyalov, 1986). More often, external forces are only a very small fraction of intracrystalline forces of mineral particles and are thus incapable of causing an oriented displacement inside particles. However, these external forces are quite adequate to activate cations of bound water to form an oriented displacement. Therefore, most soil deformation is due to the relative movement between particles.

N_0 random thermal vibrations of cations in a contact zone between two soil particles are taken as an observation sample. This sample is actually a system composed of quasi-independent subsystems. One-dimensional vibration of cations of water along a contact zone between particles contributes the most to the micro-displacement of particles. Therefore, random thermal vibration of cations is idealized to one-dimensional movement of free particles in a box with the length equal to the interval distance of atoms. According to the Schrdinger wave equation (Glasstone et al., 1941), average moving speed, $\dot{x}_a$, of cations along positive x direction can be obtained after several derivations:

$$\dot{x}_a = \left(\frac{kT}{2\pi M}\right)^{0.5} \tag{3.113}$$

where: k = Boltzmann's constant = 1.3803×10^{-16} erg per mol. K
   T = absolute temperature
   M = atom mass

If a is the interval distance of cations of firmly-bound water, the number of energy barriers crossed over per unit time, N_u, is:

$$N_u = \frac{\dot{x}_a}{a} = \frac{1}{a}\left(\frac{kT}{2\pi M}\right)^{0.5} \tag{3.114}$$

In this study, a has been assumed to be a value of 2.8×10^{-10} m (2.8 A) that is the same as the distance separating atomic valleys in the surface of a silicate mineral.

The one-dimensional translation partition function, f_x, can be written as (Eyring and Powell 1944):

$$f_x = \sum \omega_x \exp\left(-\frac{E_x}{kT}\right) \tag{3.115}$$

where: E_x = one-dimensional particle energy

ω_x = number of quantum state

The product of f_x and N_u leads to vibrating frequency of atoms, V_0, in the following form:

$$V_0 = N_u f_x = \frac{kT}{h} \tag{3.116}$$

where: h = Plank's constant = 6.624×10^{-27} erg s

- Average probability and frequency for soil particle's overcoming energy barrier

When the soil body is in an equilibrium state, cations of bound water at interface between adjacent particles may cross one or several energy barriers by accidental thermal fluctuation. However, since the probability for these cations to cross energy barriers in both forward and backward directions is the same, no macro-deformation would result. In the process of crossing one energy barrier by a particle, the introduction of an activation energy, ΔF of sufficient magnitude is required. Here, ΔF is a function of contact zone and relative position between particles as well as external loads applied on the contact zone. The potential energy of a particle may be the same following the activation process, or higher or lower than its initial value.

On the basis of the theory of absolute reaction rates by Eyring (1936) and Glasstone et al. (1941), the average probability, ρ, for soil particle's overcoming energy barrier can be derived, on the basis of the Boltzmann distribution law, as follows:

$$\rho = \exp\left(-\frac{\Delta F}{NkT}\right) \tag{3.117}$$

where: N = Avogadro's number (6.02×10^{23})

If the entire soil body is analyzed, ΔF in the above equation should be the average value for all particles.

The product of the vibrating frequency kT/h (equation 3.116) and the probability ρ (equation 3.117) determines the number of barriers succeeded by a particle per unit time, namely crossing frequency υ, as follows (Mitchell 1976):

$$v = \frac{kT}{h} \exp\left(-\frac{\Delta F}{NkT}\right) \tag{3.118}$$

In the absence of directional potentials, barriers are crossed with an equal frequency in all directions, and no macro-displacement can be observed. If, however, a directed potential is applied, then the frequency in the direction of the potential and the opposite direction changes.

- Stress-strain rate relation in a shear process

To simulate a pure shear case, consider a soil body per unit volume in Fig. 3.37 in which surfaces A and C are subject to a pressure P ($P>0$), and surfaces B and D are subject to a tensile stress $-P$. The force component p_p induced by the average allocation of the pressure P to each particle in the direction of P, may be written as:

$$p_p = \frac{P}{N_a} \tag{3.119}$$

where: N_a = number of particles per unit area, and assumed constant unless particles break down or volumetric compression occurs

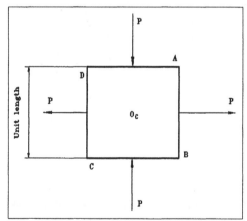

Figure 3.37 Analytical unit for shear

In a process in which a particle on a surface of soil body crosses a barrier toward the center of soil body, the pressure P exerted on the surface will exert a work equal to $p_p \times a$ (a: distance between cations of bound water), and the frequency of a particle's crossing a barrier from the surface A or C to the center can be expressed as (Mitchell, 1976):

$$\upsilon_i(A \to O_c) = \upsilon_i(C \to O_c) = V_0 \exp\left(\frac{-\Delta F / N + p_p a}{kT}\right) \tag{3.120}$$

For a tensile stress, the frequency of a particle's crossing a barrier from the surface B or D to the center can be expressed as:

$$\upsilon_i(B \to O_c) = \upsilon_i(D \to O_c) = V_0 \exp\left(\frac{-\Delta F / N + p_p a}{kT}\right) \tag{3.121}$$

where:
$V_i\ (A \to O_c)$ and $V_i\ (C \to O_c)$ = average crossing frequency of particles on
                surface A and C to the center respectively
$V_i\ (B \to O_c)$ and $V_i\ (D \to O_c)$ = average crossing frequency of particles on
                surface B and D to the center, respectively

The average translation velocity of particles on each surface is given by:

$$V_i(A \to O_c) = V_i(C \to O_c) = aV_0 \exp\left(\frac{-\Delta F / N + p_p a}{kT}\right) \tag{3.122a}$$

$$V_i(B \to O_c) = V_i(D \to O_c) = aV_0 \exp\left(\frac{-\Delta F / N + p_p a}{kT}\right) \tag{3.122b}$$

where:
$V_i\ (A \to O_c)$ and $V_i\ (C \to O_c)$ = the average translation velocity, to the center,
                of particles on surface A and C
$V_i\ (B \to O_c)$ and $V_i\ (D \to O_c)$ = the velocity, to the center, of particles on
                surface B and D, respectively

The gradients of average velocity along four shear directions are:

$$\nabla V_i(A \to B) = \nabla V_i(C \to B) = \nabla V_i(A \to D) = \nabla V_i(C \to D) =$$
$$- 2aV_0 \exp\left(-\frac{\Delta F}{NkT}\right) \sinh\left(\frac{p_p a}{kT}\right) \tag{3.123}$$

$-\nabla V$, is actually proportional to the shear strain rate of relative sliding caused by particles crossing barriers and is given by:

$$\dot{\gamma} = -\lambda_0 \nabla V_t (A \rightarrow B) = \lambda \exp\left(-\frac{\Delta F}{NkT}\right) \sinh\left(\frac{P_p a}{kT}\right) \tag{3.124}$$

where:

$$\lambda = -\frac{2\lambda_0 akT}{h} \tag{3.125}$$

$\dot{\gamma}$ = shear strain rate

λ_0 = assumed to be a constant for a specific deformation process

In the above equation, the expression $sinh(P \times a/(K \times T \times N_a))$ makes the application of this equation very complicated.

Table 3.2 Values of some parameters in equation (3.83)

Parameters	Clay	Silt	Sand
k (NmK^{-1})	1.38×10^{-23}	1.38×10^{-23}	1.38×10^{-23}
P (Nm^{-2})	10^4	10^4	10^4
a (m)	2.8×10^{-10}	2.8×10^{-10}	2.8×10^{-10}
d (m)	10^{-6}	2×10^{-5}	4×10^{-4}
N_a (m^{-2})	5.9×10^{13}	1.47×10^{11}	3.83×10^8
T (K)	300	300	300
kTN	11.46	4.58×10^3	1.76×10^6

Typical values of $P \times a/(k \times T \times N_a)$ for clay, silt and sand are listed in Table 3.2. In this table, the number of particles per unit area for clay is assumed to be 5.9×10^{13} m^{-2}, and proportional to the $1/d^2$ where d equals the diameter of the particles. The pressure P is assumed to be a value of 10 kPa which is usually encountered in agricultural production. The table indicates that the values of $P \times a/(k \times T \times N_a)$ for various types of soil are usually higher than one. Therefore, $Sh(P \times a/(k \times T \times N_a))$ can be approximately replaced by $Exp(P \times a/(k \times T \times N_a))$. equation (3.124) could then be written as:

$$\dot{\gamma} = \frac{\lambda}{2} \exp\left(-\frac{\Delta F}{NkT}\right) \exp\left(\frac{Pa}{kTN_a}\right) = \frac{\lambda}{2} \exp\left(-\frac{\Delta F}{NkT}\right) \exp\left(\frac{\tau a}{2kTN_a}\right) \tag{3.126}$$

where: τ = shear stress

Taking the natural logarithm of both sides in the above equation leads to:

$$\ln \dot{\gamma} = \ln\left(\frac{\lambda}{2}\right) - \frac{\Delta F}{NkT} + \frac{\tau a}{2kTN_a} = A + \alpha\tau \tag{3.127}$$

Where:

$$A = \ln\left(-\frac{\lambda_0 akT}{h}\right) - \frac{\Delta F}{NkT} \tag{3.128a}$$

$$\alpha = \frac{a}{2kTN_a} \tag{3.128b}$$

The semi-logarithmic relation between stress and strain rate represented by equation (3.127) was consistent with experimental results on a variety of engineering soils, such as Osaka alluvial clay (Murayama and Shibata, 1958), undisturbed London clay (Bishop, 1966), normally consolidated, overconsolidated or remolded San Francisco bay mud (Mitchell et al., 1968), Saturated Illite (Singh and Mitchell, 1968), frozen soils (Andersland and Akili, 1967), and so on.

As to agricultural soils, equation (3.127) is consistent with the results of both constant stress and constant strain rate tests. In a constant stress test, external loads and environmental temperature are constant, while in a constant strain rate test with a constant environmental temperature, strain rate of soil specimens is assumed to be constant. The temperature change in soil specimens during a deformation process is neglected.

In the case of constant external stresses and environmental temperature, there is only one variant ΔF on the right-hand side of the expression of parameter A in equation (3.128a), and ΔF is mainly related to the orientation of soil particles. Similarly, there is only one variant N_a on the right-hand side of the expression of parameter α in equation (3.128b), and N_a is also mainly dependent on the orientation of soil particles if particles are seldom broken down.

Maximyak (1968) once conducted a series of creep tests on identical specimens of clay, and focused his attention on the changes in micro-structure at various stages of deformation. According to his experimental data, the orientation of soil particles at different loads but for the same loading period appeared to be roughly the same. Therefore, it may be assumed that in a first approximation the orientation of soil particles is not influenced directly by the applied stresses and is governed, mainly, by the duration of deformation. On the basis of this assumption and the discussion in the last paragraph, it is reasonable to expect that the parameter A and α in equations (3.128a) and (3.128b) are mainly controlled by loading time t, and for any arbitrary t, $\ln \dot{\gamma}$ should have a linear relation with τ.

For unconfined tests, Fig. 3.38 illustrates experimental results of light loam. At different loading time t, logarithm of axial strain rate has a linear relation with axial stress. Fig. 3.39 shows the results of confined tests of silty clay. The linear regression results of equation (3.127) for each line in Figs. 3.38 and 3.39 are listed in Table 3.3 which indicates that logarithm of axial strain rate has a linear relation with shear stress.

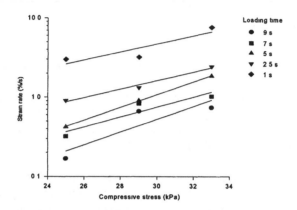

Figure 3.38   Results of unconfined compression test at constant stress (light loam, saturation degree= 98.2 %, dry density = 1600 kg m^{-3}).

Table 3.3   Regression results of equation (3.127) for constant stress tests

Figure No.	Loading time (s)	A (ln (% s^{-1}))	α (ln(% s^{-1}) kPa^{-1})	R^2
3.38	1	-6.204	0.185	0.802
3.38	2.5	-4.630	0.145	0.877
3.38	5	-5.470	0.185	0.999
3.38	7	-3.222	0.123	0.981
3.38	9	-1.964	0.117	0.804
3.39a	5	-3.684	0.0123	0.945
3.39a	5.7	-2.919	0.0121	0.955
3.39a	6.5	-2.378	0.0121	0.950
3.39a	7.3	-1.991	0.0113	0.973
3.39a	8	-1.974	0.0129	0.972
3.39b	5	-3.169	0.0126	0.916
3.39b	5.7	-2.616	0.0121	0.921
3.39b	6.5	-2.094	0.0119	0.928
3.39b	7.3	-1.789	0.0122	0.947
3.39b	8	-1.561	0.0119	0.959

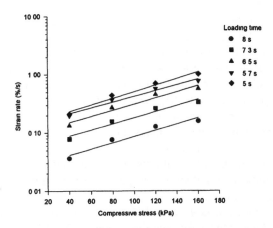

(a) clay, moisture content = 42%, dry density = 1220 kg m^{-3}

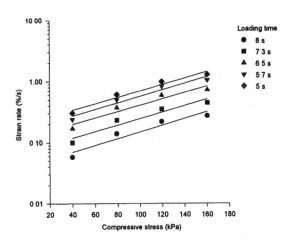

(b) clay, moisture content = 46%, dry density = 1180 kg m^{-3}

Figure 3.39 Results of confined compression test at constant stresses.

In the cases of constant strain rate tests, all parameters on the right-hand sides in equation (3.128) are constants except for ΔF and N_a. The change of ΔF and N_a mainly depends on the change in soil structure. For a constant strain-rate test, strain is proportional to loading time and thus appears to be the most useful single parameter for characterizing the change in soil structure, ΔF and N_a. Therefore, at an arbitrary strain $\ln\dot{\gamma}$ should have a linear relation with τ.

The experimental stress-strain rate relations of unconfined triaxial tests with light loam and silt clay are illustrated in Fig. 3.40. In this figure, strain of soil specimens ranges from 1% to 8%. The results of confined triaxial tests with saturated light loam are shown in Fig. 3.41 in which each straight line represents a stress-strain rate relation for a certain strain.

The shear strength-strain rate relation is a special case of shear stress-strain rate relation. The results of triaxial tests under confined conditions are shown in Fig. 3.42 thus indicating that shear strength has a linear relation with logarithm of axial strain rate for clay.

The values of parameters A, α and R^2 for each line in Figs. 3.40 and 3.41 are listed in Table 3.4 and those for each line in Fig. 3.42 are listed in Table 3.5. An approximate linear relation exists in all cases.

The applications of equation (3.128) include mainly: (1) a stress-strain-time model at a constant stress as introduced below, and (2) rate-dependent non-linear elastic model, which is introduced in Section 3.2.4, or rate-dependent elasto-plastic models.

- ### Stress-strain-time relation at constant external loads

The stress-strain-time relation at constant external loads is useful for describing the deformation of soil creep. Since strain rate $\dot{\varepsilon}$ is a function of stress σ and loading time t in a constant stress test, if we extend shear strain rate, $\dot{\gamma}$, and shear stress, τ, to any general strain rate, $\dot{\varepsilon}$, and stress, σ, which are proportional to $\dot{\gamma}$ and τ respectively, the semi-logarithmic relation in equation (3.127) can be transformed into:

$$\ln\dot{\varepsilon}(t,\sigma) = \ln\dot{\varepsilon}(t,\sigma_0) + \alpha\,\sigma \tag{3.129}$$

where: $\dot{\varepsilon}(t,\sigma_0)$ = a fictitious value of strain rate at $\sigma = 0$, a function of loading time, t;

α = value of slope of the linear portion on the logarithmic strain rate-stress plot

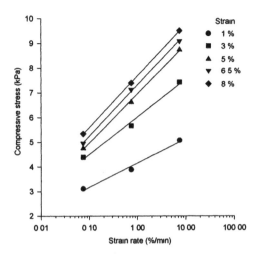

(a) light loam, saturation degree = 97.7 %, dry density = 1600 kg m^{-3}

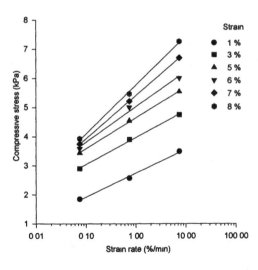

(b) silty clay, saturation degree = 100 %, dry density = 1400 kg m^{-3}

Figure 3.40 Results of unconfined compression tests at constant strain rates.

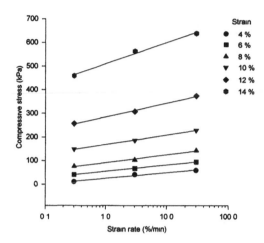

(a) light loam, saturation degree = 100 %, dry density = 1480 kg m^{-3}

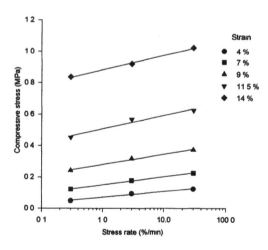

(b) light loam, saturation degree = 98.2 %, dry density = 1600 kg m^{-3}

Figure 3.41   Results of confined compression test at constant strain rate.

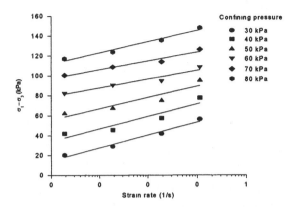

Figure 3.42 Relation between shear difference and strain rate with a clay soil.

Table 3.4 Regression results of equation (3.127) for constant strain-rate tests

Figure No.	Loading time (s)	A (ln (% s^{-1}))	α (ln(% s^{-1}) kPa^{-1})	R^2
3.40a	1	-9.62	2.31	0.985
3.40a	3	-9.08	1.51	0.991
3.40a	5	-8.03	1.16	0.999
3.40a	6.5	-8.15	1.14	0.999
3.40a	8	-8.50	1.11	0.999
3.40b	1	-7.05	2.79	0.975
3.40b	3	-9.65	2.49	0.879
3.40b	5	-13.97	2.19	0.867
3.40b	6	-23.43	1.89	0.933
3.40b	7	-28.74	1.56	0.952
3.40b	8	-26.87	1.38	0.969
3.41a	4	-2.41	0.096	0.975
3.41a	6	-4.75	0.087	0.999
3.41a	8	-6.29	0.069	0.989
3.41a	10	-9.68	0.058	0.999
3.41a	12	-11.25	0.040	0.995
3.41a	14	-13.01	0.026	0.991
3.41b	4	-4.27	0.061	0.981
3.41b	7	-6.76	0.045	0.997
3.41b	9	-9.81	0.035	0.992
3.41b	11.5	-12.98	0.026	0.961
3.41b	14	-21.85	0.025	0.996

Table 3.5 Regression results of equation (3.127) for shear difference versus strain rate relation

Figure No.	Confining pressure (kPa)	A (ln (% s^{-1}))	α (ln(% s^{-1}) kPa^{-1})	R^2
3.42a	30	-7.05	0.17	0.975
3.42a	40	-9.65	0.16	0.878
3.42a	50	-13.97	0.18	0.867
3.42a	60	-23.43	0.24	0.925
3.42a	70	-28.74	0.25	0.952
3.42a	80	-26.87	0.20	0.969

Based on the experimental studies by Shen and Yu (1989), the strain rate-time relation for agricultural soils can be expressed by:

$$\ln \dot{\varepsilon} = B - m \ln t \tag{3.130}$$

where: t = loading time
   B = constant
   m = the value of slope of the linear portion on the logarithmic strain rate-logarithmic time plot

An approximately log-log relation exists in all cases, and m is usually not equal to 1 but varies between 0.5 and 2.3. Fig. 3.43 shows a typical result of an unconfined compression test at constant stresses.

Also, because ε is a function of stress σ and loading time t, the above equation can be rewritten as:

$$\ln \dot{\varepsilon}(t,\sigma) = \ln \dot{\varepsilon}(t_1,\sigma) - m \ln\left(\frac{t}{t_1}\right) \tag{3.131}$$

where: $\dot{\varepsilon}(t_1,\sigma)$ = value of strain rate at unit time, a function of stress, σ. According to the study by Singh and Mitchell (1968), eliminating $\dot{\varepsilon}(t,\sigma)$ from equations (3.129) and (3.131) yields:

$$\ln \dot{\varepsilon}(t_1,\sigma) - m \ln\left(\frac{t}{t_1}\right) = \ln \varepsilon(t,\sigma_0) + \alpha \sigma \tag{3.132}$$

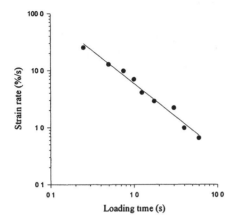

Figure 3.43 Typical results of an unconfined compression test at constant stresses.

For the case of $\sigma = 0$, the above equation can be rewritten as:

$$\ln \dot{\varepsilon}(t, \sigma_0) = \ln \varepsilon(t_1, \sigma_0) - m \ln\left(\frac{t}{t_1}\right) \tag{3.133}$$

Let $c = \dot{\varepsilon}(t_1, \sigma_0)$, the substitution of equation (3.133) into equation (3.129) leads to:

$$\dot{\varepsilon} = c \exp(\alpha \sigma)\left(\frac{t_1}{t}\right)^m \tag{3.134}$$

Based on the experimental results (Shen and Yu, 1989), it can be approximately assumed that m is not exactly equal to 1. Consequently, integration of equation (3.134) yields:

$$\varepsilon = \frac{c t_1^m}{(1-m)} \exp(\alpha \sigma) t^{(1-m)} + const$$
$$= a_1 + a_2 \exp(a_3 \sigma) t^{a_4} \tag{3.135}$$

where: a_1 through a_4 = material constants

The primary advantage of the above equation over viscoelastic models is that it includes the non-linearity of a soil. But, it is very difficult to use analytically (Christian and Desai, 1977).

4

SIMULATION OF SOIL-METAL INTERFACES

Adhesion and friction exist at every interface between the soil and a cutting tool. In some cases, they may influence the soil-tool interaction remarkably, because the adhesion c_a and external frictional angle δ at a soil-tool interface are generally different from cohesion c and internal frictional angle ϕ of the corresponding soil body, respectively. For an accurate finite element simulation, the interface should be simulated using a special type of element rather than simply using an ordinary soil element. Several commonly-used interface elements will be introduced in this chapter.

4.1 JOINT ELEMENT MODELING

Joint element is a kind of interface element degraded from a rectangular element in two dimensional cases and a cubic element in three dimensional cases. Three typical variations are presented in the following.

♦ Goodman Model
Goodman et al. (1968) proposed a 2-dimensional joint element for simulating the jointed rock interface (Ref. [1]). Later, Yong and Hanna (1977) applied it in describing the discontinuity at an interface between soil and cutting blade and a predefined cutting surface inside soil body (Ref. [2]).

The configuration of a joint element proposed by Goodman et al. is illustrated in Fig. 4.1. This element has four nodes with the nodal point pairs (1, 4) and (2, 3) having the same coordinates initially, which means that its initial width in y

direction is zero. Let L denote the length of the element in x direction, and right-hand superscripts T and B associated to a variable denote that the variable belongs to a node on the top and bottom lines, respectively.

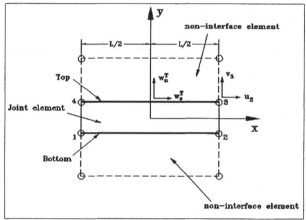

Figure 4.1 Joint element with width = 0 proposed by Goodman et al. (From Proc. Am. Soc. Civ. Eng. Soil Mechanics and Foundation , R.E. Goodman et al., 1968. Reproduced with permission of American Society of Civil Engineers, New York, NY.)

The displacement column matrix, **w**, can be expressed in terms of the nodal displacement column matrix, **u**, by a linear interpolation as follows:

$$\begin{bmatrix} w_s^B \\ w_n^B \end{bmatrix} = \frac{1}{2} \begin{bmatrix} 1-\dfrac{2x}{L} & 0 & 1+\dfrac{2x}{L} & 0 \\ 0 & 1-\dfrac{2x}{L} & 0 & 1+\dfrac{2x}{L} \end{bmatrix} \begin{bmatrix} u_1 \\ v_1 \\ u_2 \\ v_2 \end{bmatrix} \tag{4.1}$$

$$\begin{bmatrix} w_s^T \\ w_n^T \end{bmatrix} = \frac{1}{2} \begin{bmatrix} 1+\dfrac{2x}{L} & 0 & 1-\dfrac{2x}{L} & 0 \\ 0 & 1+\dfrac{2x}{L} & 0 & 1-\dfrac{2x}{L} \end{bmatrix} \begin{bmatrix} u_3 \\ v_3 \\ u_4 \\ v_4 \end{bmatrix} \tag{4.2}$$

where: u_i and v_i = displacements at node i (=1, 2, 3, 4) in the tangential and normal directions, respectively

w_s and w_n = displacements on either top or bottom of the element in the x and y directions

The relative displacement, $\Delta \mathbf{w}$, in the joint element is obtained by subtracting equation (4.1) from equation (4.2):

$$\Delta \mathbf{w} = \begin{bmatrix} w_s^T - w_s^B \\ w_n^T - w_n^B \end{bmatrix} = \frac{1}{2} \begin{bmatrix} -A & 0 & -B & 0 & B & 0 & A & 0 \\ 0 & -A & 0 & -B & 0 & B & 0 & A \end{bmatrix} \begin{bmatrix} u_1 \\ v_1 \\ u_2 \\ v_2 \\ u_3 \\ v_3 \\ u_4 \\ v_4 \end{bmatrix} \qquad (4.3)$$

where: $A = 1 - \dfrac{2x}{L}$

$B = 1 + \dfrac{2x}{L}$

To derive the element stiffness matrix, a stored energy, Ψ, caused by the applied forces per unit length acting through a virtual displacement $\Delta^{\bullet}\mathbf{w}$ is considered below:

$$\Psi = \frac{1}{2} \int_{-L/2}^{L/2} (\Delta^{\bullet}\mathbf{w})^T \mathbf{P} dx \qquad (4.4)$$

where: $\Delta^{\bullet}\mathbf{w}$ and $\mathbf{P}$ = a virtual relative displacement column matrix and a column matrix of element force per unit length, respectively. They are given by:

$$\Delta^{\bullet}\mathbf{w} = \begin{bmatrix} {}^{\bullet}w_s^T - {}^{\bullet}w_s^B \\ {}^{\bullet}w_n^T - {}^{\bullet}w_n^B \end{bmatrix} \qquad (4.5)$$

and

$$\mathbf{P} = \begin{bmatrix} P_s \\ P_n \end{bmatrix} \qquad (4.6)$$

Goodman et al. considered that the element force matrix, **P**, might be expressed in terms of a product of element stiffness and displacement:

$$\mathbf{P} = \mathbf{k_d}\, \Delta^{\!*} \mathbf{w} \tag{4.7}$$

where: $\mathbf{k_d}$ = a matrix of diagonal stiffness per unit length and of the following simplest form:

$$\mathbf{k_d} = \begin{bmatrix} k_s & 0 \\ 0 & k_n \end{bmatrix} \tag{4.8}$$

Substitution of equation (4.7) into equation (4.4) leads to:

$$\Psi = \frac{1}{2} \int_{-L/2}^{L/2} (\Delta^{\!*}\mathbf{w})^T \mathbf{k_d} \Delta^{\!*} \mathbf{w}\, dx \tag{4.9}$$

The relation in equation (4.3) can be symbolically rewritten as:

$$\Delta\mathbf{w} = \frac{1}{2}\mathbf{D}\,\mathbf{u} \tag{4.10}$$

where: **D** and **u** = the first and second matrix on the right-hand side in equation (4.3), respectively. Application of the relation in equation (4.10) to the case of the virtual displacement, $^{\!*}\mathbf{u}$, leads to:

$$\Delta^{\!*}\mathbf{w} = \frac{1}{2}\mathbf{D}\,^{\!*}\mathbf{u} \tag{4.11}$$

Substitution of equation (4.11) into equation (4.9) yields:

$$\Psi = \frac{1}{2} \int_{-L/2}^{L/2} \frac{1}{4} (^{\!*}\mathbf{u})^T \mathbf{D}^T \mathbf{k_d} \mathbf{D}\,^{\!*}\mathbf{u}\, dx \tag{4.12}$$

where: $\mathbf{D}^T \mathbf{k_d} \mathbf{D}$ is given by:

$$
\mathbf{D}^T \mathbf{k_d} \mathbf{D} =
\begin{bmatrix}
-A & 0 \\
0 & -A \\
-B & 0 \\
0 & -B \\
B & 0 \\
0 & B \\
A & 0 \\
0 & A
\end{bmatrix}
\begin{bmatrix}
k_s & 0 \\
0 & k_n
\end{bmatrix}
\begin{bmatrix}
-A & 0 & -B & 0 & B & 0 & A & 0 \\
0 & -A & 0 & -B & 0 & B & 0 & A
\end{bmatrix}
\quad (4.13)
$$

Performing matrix multiplication of the above equation results in:

$$
\mathbf{D}^T \mathbf{k_d} \mathbf{D} =
\begin{bmatrix}
A^2 k_s & 0 & ABk_s & 0 & -ABk_s & 0 & -A^2 k_s & 0 \\
0 & A^2 k_n & 0 & ABk_n & 0 & -ABk_n & 0 & -A^2 k_n \\
ABk_s & 0 & B^2 k_s & 0 & -B^2 k_s & 0 & -ABk_s & 0 \\
0 & ABk_n & 0 & B^2 k_n & 0 & -B^2 k_n & 0 & -ABk_n \\
-ABk_s & 0 & -B^2 k_s & 0 & B^2 k_s & 0 & ABk_s & 0 \\
0 & -ABk_n & 0 & -B^2 k_n & 0 & B^2 k_n & 0 & ABk_n \\
-A^2 k_s & 0 & -ABk_s & 0 & ABk_s & 0 & A^2 k_s & 0 \\
0 & -A^2 k_n & 0 & -ABk_n & 0 & ABk_n & 0 & A^2 k_n
\end{bmatrix}
$$
$$(4.14)$$

Goodman et al. considered that the stored energy, Ψ, in equation (4.12) could be expressed in terms of the matrix of joint element stiffness per unit length, $\mathbf{k}$, as follows:

$$
\Psi = \frac{1}{2} L(^\bullet \mathbf{u})^T \mathbf{k} \, ^\bullet \mathbf{u}
\tag{4.15}
$$

Substituting equation (4.14) into equation (4.12), performing the integration and then comparing the result with equation (4.15) will lead to the expression of $\mathbf{k}$ as follows:

$$
\mathbf{k} = \frac{1}{6}
\begin{bmatrix}
2k_s & 0 & k_s & 0 & -k_s & 0 & -2k_s & 0 \\
0 & 2k_n & 0 & k_n & 0 & -k_n & 0 & -2k_n \\
k_s & 0 & 2k_s & 0 & -2k_s & 0 & -k_s & 0 \\
0 & k_n & 0 & 2k_n & 0 & -2k_n & 0 & -k_n \\
-k_s & 0 & -2k_s & 0 & 2k_s & 0 & k_s & 0 \\
0 & -k_n & 0 & -2k_n & 0 & 2k_n & 0 & k_n \\
-2k_s & 0 & -k_s & 0 & k_s & 0 & 2k_s & 0 \\
0 & -2k_n & 0 & -k_n & 0 & k_n & 0 & 2k_n
\end{bmatrix}
\tag{4.16}
$$

The last step to derive the stiffness matrix of a joint element is to transform it from the local coordinate system to a global coordinate system. For the coordinate systems shown in Fig. 4.2, the transformation matrix operation is of the form:

$$
\begin{bmatrix} x \\ y \end{bmatrix} =
\begin{bmatrix} \cos\theta & \sin\theta \\ -\sin\theta & \cos\theta \end{bmatrix}
\begin{bmatrix} X \\ Y \end{bmatrix}
\tag{4.17}
$$

where: x and y = local coordinates in the tangential and normal directions, respectively

X and Y = global coordinates

After an element stiffness matrix is transformed to a global coordinate system, it is ready for being assembled to form a global stiffness matrix, $\mathbf{K}$.

One shortcoming associated to this element is that it may cause the penetration of adjacent blocks of continuum elements into each other, which is not possible in reality.

♦ Ghaboussi Model

Ghaboussi et al. (1973) made some improvement in geometric description of a joint element. Relative displacement is viewed as the independent degree-of-freedom of the new joint element, as shown in Fig. 4.3. The displacements on the top side of the element are transformed into the relative displacements between the top and bottom sides as follows:

$$
\begin{aligned}
u_4^T &= u_1^B + \Delta u_4; \quad v_4^T = v_1^B + \Delta v_4 \\
u_3^T &= u_2^B + \Delta u_3; \quad v_3^T = v_2^B + \Delta v_3
\end{aligned}
\tag{4.18}
$$

where, superscripts, T and B, denote that the corresponding variables belong to a node on the top and bottom sides, respectively. As in Fig. 4.3, the displacements at nodes on the top side are transformed to relative quantities, while the displacements on the bottom side remain their original values.

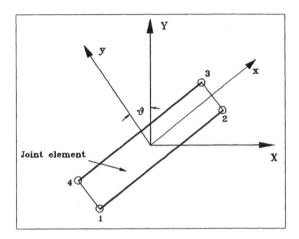

Figure 4.2 A local coordinate system x-y of a joint element and a global coordinate system X-Y.

Considering that the pairs of nodes (1, 4) and (2, 3) are very close, it is permissible to use only two nodes (i, j) and a thickness t to represent the entire geometry of the element, as illustrated in Fig. 4.4. The relative displacements at nodes i and j are considered equal to those at nodes 4 and 3, that is,

$$\Delta u_i = \Delta u_4; \quad \Delta v_i = \Delta v_4$$
$$\Delta u_j = \Delta u_3; \quad \Delta v_j = \Delta v_3 \tag{4.19}$$

A simple coordinate rotation in Fig. 4.4 can be performed on the relative displacements at nodes i and j as follows:

$$\begin{bmatrix} \Delta u_{si} \\ \Delta v_{ni} \\ \Delta u_{sj} \\ \Delta v_{nj} \end{bmatrix} = \begin{bmatrix} \cos\theta & \sin\theta & 0 & 0 \\ -\sin\theta & \cos\theta & 0 & 0 \\ 0 & 0 & \cos\theta & \sin\theta \\ 0 & 0 & -\sin\theta & \cos\theta \end{bmatrix} \begin{bmatrix} \Delta u_i \\ \Delta v_i \\ \Delta u_j \\ \Delta v_j \end{bmatrix} \tag{4.20}$$

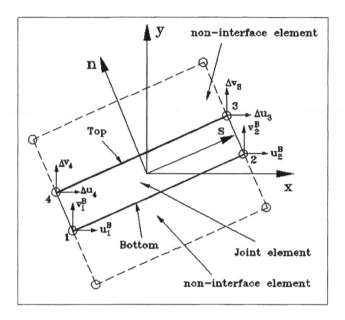

Figure 4.3 Geometry of joint element proposed by Ghaboussi et al.

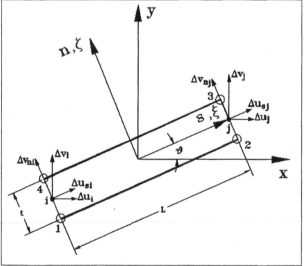

Figure 4.4 Coordinate systems of a two-dimensional joint element.

The relative tangential and normal displacements, Δu_s and Δv_n, inside the element are assumed to vary linearly as follows:

$$\Delta u_s = h_i \Delta u_{si} + h_j \Delta u_{sj}$$
$$\Delta v_n = h_i \Delta v_{ni} + h_j \Delta v_{nj}$$

(4.21)

where: h_i and h_j = linear interpolation functions expressed by:

$$h_i = \frac{1}{2}(1-\xi); \quad h_j = \frac{1}{2}(1+\xi)$$

(4.22)

where, ξ as well as ζ in Fig. 4.4 represent a non-dimensional coordinate system with the following definition:

$$\xi = \frac{2}{L}s; \quad \zeta = \frac{2}{t}n$$

(4.23)

It is assumed that there are only two strain components in the joint element with the following definitions:

$$\varepsilon_n = \frac{1}{t}\Delta v_n; \quad \varepsilon_s = \frac{1}{t}\Delta u_s$$

(4.24)

Substitution of equations (4.21) and (4.22) into equation (4.24) leads to:

$$\begin{bmatrix} \varepsilon_n \\ \varepsilon_s \end{bmatrix} = \frac{1}{2t}\begin{bmatrix} 1-\xi & 0 & 1+\xi & 0 \\ 0 & 1-\xi & 0 & 1+\xi \end{bmatrix}\begin{bmatrix} \Delta v_{ni} \\ \Delta u_{si} \\ \Delta v_{nj} \\ \Delta u_{sj} \end{bmatrix}$$

(4.25)

$$\varepsilon = \mathbf{B}\Delta\mathbf{u}$$

The stresses and strains are related using the following equation:

$$\begin{bmatrix} \sigma_n \\ \sigma_s \end{bmatrix} = \begin{bmatrix} C_{nn} & C_{ns} \\ C_{sn} & C_{ss} \end{bmatrix}\begin{bmatrix} \varepsilon_n \\ \varepsilon_s \end{bmatrix}$$

(4.26)

$$\sigma = \mathbf{C}\varepsilon$$

The element stiffness matrix in the *n-s* local coordinate system is given by:

$$\mathbf{k}_{ns} = \int_{vol} \mathbf{B}^T \mathbf{C} \mathbf{B} dV \tag{4.27}$$

Transforming it to a global *x-y* coordinate system results in:

$$\mathbf{k} = \mathbf{T}^T \mathbf{k}_{ns} \mathbf{T} \tag{4.28}$$

where, **T** is the transformation matrix consisting of the direction cosines which result in:

$$\mathbf{k} = \frac{L}{6t} \begin{bmatrix} 2(A_1 - 2B_1) & 2(A_3 + B_2) & A_1 - 2B_1 & A_3 + B_2 \\ & 2(A_2 + B_1) & A_3 + B_2 & A_2 + 2B_1 \\ & & 2(A_1 - 2B_1) & 2(A_3 + B_2) \\ symmetric & & & 2(A_2 + 2B_2) \end{bmatrix} \tag{4.29}$$

where:

$$A_1 = C_{ss}a^2 + C_{nn}b^2; \quad B_1 = C_{ns}ab$$
$$A_2 = C_{ss}b^2 + C_{nn}a^2; \quad B_2 = C_{ns}(a^2 - b^2)$$
$$A_3 = (C_{nn} - C_{ss})ab$$

and

$$a = \frac{1}{L}(x_j - x_i); \quad b = \frac{1}{L}(y_j - y_i)$$

The different forms of the matrix **C** in equation (4.26) correspond to specified material properties of the joint element. The most simplest **C** is of the diagonal form as follows:

$$\begin{bmatrix} \sigma_n \\ \sigma_s \end{bmatrix} = \begin{bmatrix} C_{nn} & 0 \\ 0 & C_{ss} \end{bmatrix} \begin{bmatrix} \varepsilon_n \\ \varepsilon_s \end{bmatrix} \tag{4.30}$$

which implies that no volume change would occur due to shearing strains, and thus the shear and the normal components are uncoupled. An element of this form of matrix **C** is called the non-dilatent joint element.

Both C_{nn} and C_{ss} in equation (4.30) should be non-linear functions of strain and stress levels. When an interface is in a tension state, i.e. $\varepsilon_n \geq 0$, $C_{nn} = C_{ss}$

= 0. If an interface is in a compression state, $C_{nn} = E_f$ which may be of a very large magnitude, and C_{ss} may be of the following form:

$$C_{ss} = \begin{cases} G & \sigma_s < c_a + \sigma_n \tan\delta \quad \text{(elastic)} \\ 0 & \sigma_s = c_a + \sigma_n \tan\delta \quad \text{(plastic)} \end{cases}$$

where: c_a = adhesion at an interface
δ = external frictional angle at an interface

The **C** in equation (4.26) can be of a more complicated form than that in equation (4.30)to simulate some complex behavior such as dilatancy. But, for the interface between soil and tool, it cannot be seen that there is any benefit to do so.

One possible problem associated with a joint element is that when shear stress exceeds the limitation constrained by the friction criterion at an interface, either reducing the shear stress or increasing the normal stress or both can be tried to meet the requirement of maximum shear stress allowed by the friction criterion. This may lead to a different solution to an identical problem, and therefore cause some uncertainty.

◆ Zienkiewicz Model

Zienkiewicz et al. (1970) advocated the use of isoparametric joint element. The same coordinate transformation as shown in Fig. 4.4 and equation (4.23) is used. The displacement mode of the element is expressed by the following linear relation:

$$\begin{bmatrix} u \\ v \end{bmatrix} = [N(\xi,\zeta)]\delta \tag{4.31}$$

where: δ = a column matrix of four sets of nodal displacements
$[N(\xi,\zeta)]$ = a shape function matrix with a typical form:

$$N_i = \overline{N}_i \overline{\overline{N}}_i$$
$$\overline{N}_i = \frac{1}{2}(1-\xi_0) \quad \overline{\overline{N}}_i = \frac{1}{2}(1-\eta_0) \tag{4.32}$$
$$\xi_0 = \xi\xi_i \quad \eta_0 = \eta\eta_i$$

Details of derivation of this element   follow a standard procedure for isoparametric elements (Zienkiewicz, 1989). Some particular features of the element are:

(a) The strain components are allowed to vary linearly inside the element in both ξ and ζ directions which corresponds to the local coordinate system.

(b) Since the thickness t is very small, it is sufficient to carry out the numerical integration at only two Gauss points on the ξ axis. The stresses and strains at the centroid would consist of adequate information to the system.

One potential disadvantage of this element in some cases is that the stiffness matrix may become ill-conditioning caused by very large off-diagonal entries or very small diagonal entries, leading to some numerical difficulties.

◆   Wilson Model

Wilson (1975) formalized the procedure to develop a three-dimensional isoparametric joint element. The joint element is expressed by four nodes as shown in Fig. 4.5. The unknown variables include three components of displacements u, v, w at each standard node and u_s, v_t , w_n at each interface node. Δu_s, Δv_t and Δw_n refers to the increments of u_s, v_t , w_n, respectively.

In order to establish the stiffness matrix for an interface element, the displacement mode, i.e., the relation between displacements (u_s, v_t and w_n) in the element and the displacements at the eight nodes needs to be chosen. Using the standard shape functions for an eight-node element, the following linear functions are usually used:

$$u_s = \sum_{i=1}^{8} H_i u_{si} \qquad v_t = \sum_{i=1}^{8} H_i v_{ti} \qquad w_n = \sum_{i=1}^{8} H_i w_{ni} \qquad (4.33)$$

The displacements at nodes 5 to 8 in Fig. 4.6 can be expressed by:

$$u_{si} = u_{s(i-4)} + \Delta u_{s(i-4)}$$
$$v_{ti} = v_{t(i-4)} + \Delta v_{t(i-4)} \qquad (i = 5...8) \qquad (4.34a)$$
$$w_{ni} = w_{n(i-4)} + \Delta w_{n(i-4)}$$

and

$$H_i(\xi,\eta,\zeta) = \frac{1-\zeta}{2} N_i(\xi,\eta) \qquad (i = 1...4)$$

$$\qquad\qquad\qquad\qquad\qquad\qquad\qquad\qquad (4.34b)$$

$$H_i(\xi,\eta,\zeta) = \frac{1+\zeta}{2} N_i(\xi,\eta) \qquad (i = 5...8)$$

where:

$$N_i(\xi, \eta) = N_{i-4}(\xi, \eta) \qquad (i = 5...8)$$

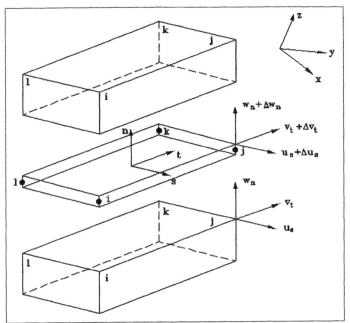

Figure 4.5 Three-dimensional joint element proposed by Wilson. (From *Finite Elements in Geomechanics* (edited by G. Gudehus), E.L. Wilson, 1977. Reproduced with permission of John Wiley & Sons, Ltd, West Sussex, England.)

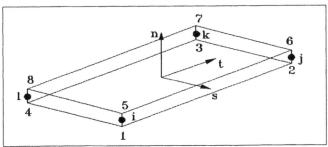

Figure 4.6 Local reference system for a joint element.

Considering equation (4.34), equation (4.33) can be rewritten as:

$$u_s = \sum_{t=1}^{4} \left(\frac{1-\zeta}{2} N_t u_{st} + \frac{1+\zeta}{2} N_t u_{st} + \frac{1+\zeta}{2} N_t \Delta u_{st} \right)$$

$$v_t = \sum_{t=1}^{4} \left(\frac{1-\zeta}{2} N_t v_{tt} + \frac{1+\zeta}{2} N_t v_{tt} + \frac{1+\zeta}{2} N_t \Delta v_{tt} \right) \qquad (4.35)$$

$$w_n = \sum_{t=1}^{4} \left(\frac{1-\zeta}{2} N_t w_m + \frac{1+\zeta}{2} N_t w_m + \frac{1+\zeta}{2} N_t \Delta w_m \right)$$

With $(1+\zeta)/2 = t/h$ in the global coordinate system, equation (4.35) is further simplified to:

$$u_s = \sum_{t=1}^{4} \left(N_t u_{st} + \frac{h}{t} N_t \Delta u_{st} \right)$$

$$v_t = \sum_{t=1}^{4} \left(N_t v_{tt} + \frac{h}{t} N_t \Delta v_{tt} \right) \qquad (4.36)$$

$$w_n = \sum_{t=1}^{4} \left(N_t w_m + \frac{h}{t} N_t \Delta w_m \right)$$

If all strains are assumed to be constant in the thickness direction n, the strains in the element are defined by:

$$\varepsilon_{ss} = \frac{\partial u_s}{\partial s} = \sum N_{t,s} u_{st}$$

$$\varepsilon_{tt} = \frac{\partial u_t}{\partial t} = \sum N_{t,t} u_{tt}$$

$$\varepsilon_{nn} = \frac{\partial u_n}{\partial n} = \sum N_{t,n} u_{ni}$$

$$\varepsilon_{st} = \frac{\partial u_s}{\partial t} + \frac{\partial u_t}{\partial s} = \sum N_{t,t} u_{st} + N_{t,s} u_{tt} \qquad (4.37)$$

$$\varepsilon_{sn} = \frac{\partial u_s}{\partial n} + \frac{\partial u_n}{\partial s} = \sum \frac{1}{h} N_t \Delta u_{st} + N_{t,s} u_m$$

$$\varepsilon_{tn} = \frac{\partial u_t}{\partial n} + \frac{\partial u_n}{\partial t} = \sum \frac{1}{h} N_t \Delta u_{tt} + N_{t,t} u_m$$

Some of the strain components in equation (4.37) can be omitted in many cases. The remaining strain components should be consistent with a set of material properties selected. One advantage of using shape functions is that the computer implementation of the element is very similar to that of the isoparametric solid element.

4.2 FRICTION ELEMENT MODELING

Hou and Liu (1986) used a rod-like friction element in simulating a soil-tool interface. The element has two nodes with a length L, as shown in Fig. 4.7. The node i is one node adjacent to a soil element, while node j is considered as one point on the tool surface.

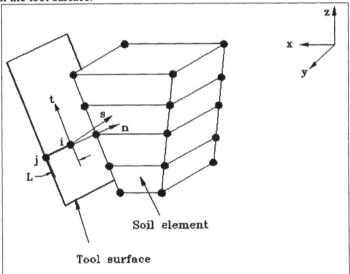

Figure 4.7 Friction element proposed by Hou and Liu. (From Transactions of Chinese Society of Agricultural Machinery, Hou, Z. et al., 1986. Reproduced with permission of Chinese Society of Agricultural Machinery, Beijing, China.)

Let P_{si}, P_{ti} and P_{ni} denote total internal nodal force components at node i exerted by its adjacent elements excluding the friction element, and P_{sj}, P_{tj} and P_{nj} the total internal nodal forces at node j. Since the magnitude of L is very small, it is permissible to assume that nodal force components at nodes i and j are equal in magnitude and opposite in direction. Therefore, for the friction element with nodes i and j, six equilibrium equations in local coordinate system *s-t-n* can be written as:

$$P_{si} + P_{sj} = 0$$
$$P_{ti} + P_{tj} = 0$$
$$P_{ni} + P_{nj} = 0$$
$$(u_{si} - u_{sj})k_s = P_{si} \qquad (4.38)$$
$$(v_{ti} - v_{tj})k_t = P_{ti}$$
$$(w_{ni} - w_{nj})k_n = P_{ni}$$

where: $u_{si}, v_{ti}, w_{ni}, u_{sj}, v_{tj}, w_{nj}$ = displacement components in directions s, t and n at nodes i and j, respectively

k_s, k_t and k_n = stiffness coefficients in directions s, t and n, respectively

The equilibrium equations in a global coordinate system *x-y-z* are given by:

$$P_{xi} + P_{xj} = 0$$
$$P_{yi} + P_{yj} = 0 \qquad (4.39)$$
$$P_{zi} + P_{zj} = 0$$

The coordinate transformation can be defined by:

$$\begin{bmatrix} s \\ t \\ n \end{bmatrix} = \begin{bmatrix} s_1 & s_2 & s_3 \\ t_1 & t_2 & t_3 \\ n_1 & n_2 & n_3 \end{bmatrix} \begin{bmatrix} x \\ y \\ z \end{bmatrix} \qquad (4.40)$$

where:

$$s_1 = \cos(s,x) \quad s_2 = \cos(s,y) \quad s_3 = \cos(s,z)$$
$$t_1 = \cos(t,x) \quad t_2 = \cos(t,y) \quad t_3 = \cos(t,z)$$
$$n_1 = \cos(n,x) \quad n_2 = \cos(n,y) \quad n_3 = \cos(n,z)$$

where: (i, j) = the angle between the directions i and j ($I = s$, t or n; $j = x$, y or z)

Considering the coordinate transformation in equation (4.40), the following displacement and force transformations are given, respectively, by:

$$
\begin{bmatrix} u_{si} \\ v_{ti} \\ w_{ni} \\ u_{sj} \\ v_{tj} \\ w_{nj} \end{bmatrix} = \begin{bmatrix} s_1 & s_2 & s_3 & 0 & 0 & 0 \\ t_1 & t_2 & t_3 & 0 & 0 & 0 \\ w_1 & w_2 & w_3 & 0 & 0 & 0 \\ 0 & 0 & 0 & s_1 & s_2 & s_3 \\ 0 & 0 & 0 & t_1 & t_2 & t_3 \\ 0 & 0 & 0 & w_1 & w_2 & w_3 \end{bmatrix} \begin{bmatrix} u_i \\ v_i \\ w_i \\ u_j \\ v_j \\ w_j \end{bmatrix} = \mathbf{T} \begin{bmatrix} u_i \\ v_i \\ w_i \\ u_j \\ v_j \\ w_j \end{bmatrix} \tag{4.41}
$$

and

$$
\begin{bmatrix} P_{si} \\ P_{ti} \\ P_{ni} \\ P_{sj} \\ P_{tj} \\ P_{nj} \end{bmatrix} = \begin{bmatrix} s_1 & s_2 & s_3 & 0 & 0 & 0 \\ t_1 & t_2 & t_3 & 0 & 0 & 0 \\ w_1 & w_2 & w_3 & 0 & 0 & 0 \\ 0 & 0 & 0 & s_1 & s_2 & s_3 \\ 0 & 0 & 0 & t_1 & t_2 & t_3 \\ 0 & 0 & 0 & w_1 & w_2 & w_3 \end{bmatrix} \begin{bmatrix} P_{xi} \\ P_{yi} \\ P_{zi} \\ P_{xj} \\ P_{yj} \\ P_{zj} \end{bmatrix} = \mathbf{T} \begin{bmatrix} P_{xi} \\ P_{yi} \\ P_{zi} \\ P_{xj} \\ P_{yj} \\ P_{zj} \end{bmatrix} \tag{4.42}
$$

Equation (4.38) can be rewritten in a matrix form:

$$
\begin{bmatrix} k_s & 0 & 0 & -k_s & 0 & 0 \\ 0 & k_t & 0 & 0 & -k_t & 0 \\ 0 & 0 & k_n & 0 & 0 & -k_n \\ -k_s & 0 & 0 & k_s & 0 & 0 \\ 0 & -k_t & 0 & 0 & k_t & 0 \\ 0 & 0 & -k_n & 0 & 0 & k_n \end{bmatrix} \begin{bmatrix} u_{si} \\ v_{ti} \\ w_{ni} \\ u_{sj} \\ v_{tj} \\ w_{nj} \end{bmatrix} = \begin{bmatrix} P_{si} \\ P_{ti} \\ P_{ni} \\ P_{sj} \\ P_{tj} \\ P_{nj} \end{bmatrix} \tag{4.43}
$$

After a coordinate transformation, it becomes:

$$
\begin{bmatrix} P_{xi} \\ P_{yi} \\ P_{zi} \\ P_{xj} \\ P_{yj} \\ P_{zj} \end{bmatrix} = \mathbf{T}^{-1} \begin{bmatrix} k_s & 0 & 0 & -k_s & 0 & 0 \\ 0 & k_t & 0 & 0 & -k_t & 0 \\ 0 & 0 & k_n & 0 & 0 & -k_n \\ -k_s & 0 & 0 & k_s & 0 & 0 \\ 0 & -k_t & 0 & 0 & k_t & 0 \\ 0 & 0 & -k_n & 0 & 0 & k_n \end{bmatrix} \mathbf{T} \begin{bmatrix} u_i \\ v_i \\ w_i \\ u_j \\ v_j \\ w_j \end{bmatrix} \tag{4.44}
$$

Therefore, the element stiffness matrix in the global coordinate system is of the form:

$$
\mathbf{k} = \mathbf{T}^{-1}
\begin{bmatrix}
k_s & 0 & 0 & -k_s & 0 & 0 \\
0 & k_t & 0 & 0 & -k_t & 0 \\
0 & 0 & k_n & 0 & 0 & -k_n \\
-k_s & 0 & 0 & k_s & 0 & 0 \\
0 & -k_t & 0 & 0 & k_t & 0 \\
0 & 0 & -k_n & 0 & 0 & k_n
\end{bmatrix}
\mathbf{T} =
\begin{bmatrix}
\mathbf{k}_{ii} & \mathbf{k}_{ij} \\
\mathbf{k}_{ji} & \mathbf{k}_{jj}
\end{bmatrix}
\tag{4.45}
$$

where:

$$
\mathbf{k}_{ii} =
\begin{bmatrix}
s_1^2 k_s + t_1^2 k_t + w_1^2 k_n & s_1 s_2 k_s + t_1 t_2 k_t + w_1 w_2 k_n & s_1 s_3 k_s + t_1 t_3 k_t + w_1 w_3 k_n \\
s_1 s_2 k_s + t_1 t_2 k_t + w_1 w_2 k_n & s_2^2 k_s + t_2^2 k_t + w_2^2 k_n & s_2 s_3 k_s + t_2 t_3 k_t + w_2 w_3 k_n \\
s_1 s_3 k_s + t_1 t_3 k_t + w_1 w_3 k_n & s_2 s_3 k_s + t_2 t_3 k_t + w_2 w_3 k_n & s_3^2 k_s + t_3^2 k_t + w_3^2 k_n
\end{bmatrix}
$$

and

$$
\mathbf{k}_{jj} = \mathbf{k}_{ii} \quad \mathbf{k}_{ji} = \mathbf{k}_{ij} = -\mathbf{k}_{ii}
$$

At any moment, the friction element may be in one of the following four states:

(1) separation state

A separation state means that a tool element already loses contact with the corresponding soil element. Its identification criterion is $\Delta w_{ni} - \Delta w_{nj} > 0$ in which Δw_{ni} and Δw_{nj} are relative displacements in direction n at nodes i and j, respectively. The result of this state is $\tilde{P}_i = \tilde{P}_j = 0$ in which $\tilde{P}_i$ and $\tilde{P}_j$ are nodal forces at nodes *i* and *j*, respectively.

(2) stick state

In a stick state, the soil and tool elements contact each other without any relative movement. The identification criterion is:

$$
\Delta w_{ni} - \Delta w_{nj} \le 0, \quad \Delta w_{si} - \Delta w_{sj} = 0 \text{ and } \Delta w_{ti} - \Delta w_{tj} = 0
$$

where: Δw_{si}, Δw_{sj}, Δw_{ti} and Δw_{tj} = relative displacements in directions s and t at nodes i and j, respectively

In this state, the nodal forces are constrained by the equilibrium equation:

$$\tilde{P}_i = -\tilde{P}_j$$

(3) slip state

In a slip state, the soil and tool elements may slide in direction s or t or both. If the slip occurs only in direction s, the identification criterion and force equilibrium equation are given by:

$$\Delta w_{ni} - \Delta w_{nj} \leq 0, \ \Delta w_{si} - \Delta w_{sj} \neq 0, \ \Delta w_{ti} - \Delta w_{tj} = 0$$
$$P_{ti} = -P_{tj}, \quad P_{ni} = -P_{nj}, \quad P_{si} = -P_{sj} = \pm f \, P_{ni}$$

where: f = friction coefficient

If the slip takes place only in direction t, two similar equations are applied as above:

$$\Delta w_{ni} - \Delta w_{nj} \leq 0, \ \Delta w_{si} - \Delta w_{sj} = 0, \ \Delta w_{ti} - \Delta w_{tj} \neq 0$$
$$P_{si} = -P_{sj}, \quad P_{ni} = -P_{nj}, \quad P_{ti} = -P_{tj} = \pm f \, P_{ni}$$

In cases where the slip exists in both directions s and t, the above two equations are changed to:

$$\Delta w_{ni} - \Delta w_{nj} \leq 0, \ \Delta w_{si} - \Delta w_{sj} \neq 0, \ \Delta w_{ti} - \Delta w_{tj} \neq 0$$
$$P_{ni} = -P_{nj}, \quad P_{si} = -P_{sj} = \pm f \, P_{ni}, \quad P_{ti} = -P_{tj} = \pm f \, P_{ni}$$

In order to avoid the penetration of a node in soil elements into a tool element, k_n should be assigned a value with a large magnitude, for example, 10^8 (m kPa). Both k_s and k_t depend upon material properties which we will address in Section 4.4.

For the case of large soil displacements, after each incremental step, node coordinates of a friction element can be corrected to form a new element, as shown in Fig. 4.8. Assume that i' is the position of node i after an incremental step. Using the new position i' as the starting point, we make a perpendicular line to the tool surface and obtain an intersection point j'. In this way, a new friction element i'-j' is generated.

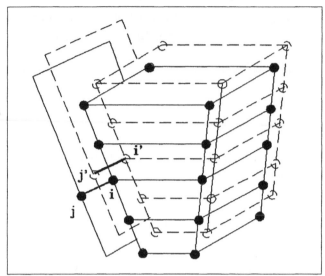

Figure 4.8 Generation of a new friction element.

The coordinates at node i after the n-th incremental step are determined by:

$$
\begin{aligned}
{}^{n}x_{i} &= {}^{0}x_{i} + {}^{n}u_{i} \\
{}^{n}y_{i} &= {}^{0}y_{i} + {}^{n}v_{i} \\
{}^{n}z_{i} &= {}^{0}z_{i} + {}^{n}w_{i}
\end{aligned}
\qquad (4.46)
$$

where: $^{0}x_{i},\ ^{0}y_{i},\ ^{0}z_{i}$ = initial coordinates at node i
           $^{n}u_{i},\ ^{n}v_{i},\ ^{n}w_{i}$ = total displacement components after the n-th step
           $^{n}x_{i},\ ^{n}y_{i},\ ^{n}z_{i}$ = new coordinates at node i after the n-th step

The new coordinates at node j after the n-th incremental step are:

$$
\begin{aligned}
{}^{n}x_{j} &= {}^{n}x_{i} + Ln_{1} \\
{}^{n}y_{j} &= {}^{n}y_{i} + Ln_{2} \\
{}^{n}z_{j} &= {}^{n}z_{i} + Ln_{3}
\end{aligned}
\qquad (4.47)
$$

The generation of a new element would not influence the forces within the element, because the internal forces within a friction element depend only upon the equivalent forces of internal stresses in adjacent non-interface elements at a specific incremental step, that is,

$$\mathbf{P}_i = \begin{bmatrix} P_{xi} \\ P_{yi} \\ P_{zi} \end{bmatrix} = \left(\sum \mathbf{k} \ \mathbf{u} \right)_i = \left(\int \mathbf{B}^T \sigma \, dV \right)_i \qquad (4.48)$$

where: $(\ )_i$ = the summation of only elements which are adjacent to node i

$\mathbf{k, u}, \sigma$ = element stiffness, displacement and stress matrices, respectively

4.3 THIN-LAYER ELEMENT MODELING

Desai et al. (1984) proposed a thin solid element to simulate interface behavior. Since the proposed element essentially represents a solid element with a small finite thickness, Desai et al. called it a "thin layer" element. Schematic views of a two- and three-dimensional thin-layer element are given in Fig. 4.9. The only difference between this element and a standard solid element is that it has a particular constitutive matrix, $\mathbf{C}^i$ such that

$$d\sigma = \mathbf{C}' d\varepsilon \qquad (4.49)$$

where: $d\sigma$ and $d\varepsilon$ = column matrices of increments of stresses and strains, respectively

$\mathbf{C}'$ is given by:

$$\mathbf{C}' = \begin{bmatrix} \mathbf{C}'_{nn} & \mathbf{C}'_{ns} \\ \mathbf{C}'_{sn} & \mathbf{C}'_{ss} \end{bmatrix} \qquad (4.50)$$

where, $\mathbf{C}'_{nn}$ and $\mathbf{C}'_{ss}$ are normal and shear components, respectively; $\mathbf{C}'_{ns}$ and $\mathbf{C}'_{sn}$ represent coupling effects. Since it is difficult to determine the coupling effects through laboratory tests, $\mathbf{C}^i_{ns}$ and $\mathbf{C}'_{sn}$ are usually considered to be composed of zero.

Desai considered that it might not be appropriate to assign an arbitrarily high value for $\mathbf{C}'_{nn}$ and proposed the following equation to calculate the normal stiffness:

$$\mathbf{C}'_{nn} = \mathbf{C}'_{nn}(\alpha'_m, \beta^g_m, \gamma^{st}_m) \qquad (4.51)$$

where, $\alpha_m^I, \beta_m^g, \gamma_m^{sl}$ $(m = 1,2,...)$ represent the properties of the interface, geological and structural elements, respectively. equation (4.51) can be written as:

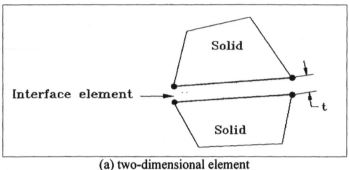

(a) two-dimensional element

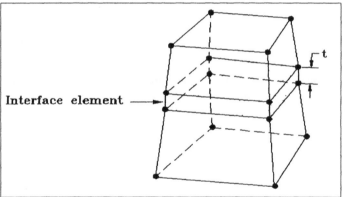

(b) three-dimensional element

Figure 4.9 Thin-layer interface element proposed by Desai et al. (From *International Journal for Numerical and Analytical Methods in Geomechanics*, Desai, C.S. et. al., 1984. Reproduced with permission of John Wiley & Sons Limited, Sussex, U.K.)

$$\overline{C}_{nn}^I = \lambda_1 C_{nn}^I + \lambda_2 C_{nn}^g + \lambda_3 C_{nn}^{st} \tag{4.52}$$

where: $\overline{C}_{nn}^I$ = parameter that refers to the normal behavior of a thin-layer element

λ_1, λ_2 and λ_3 = participation factors with the sum equal to 1. For most cases, it is assumed that $\lambda_1 = 1$ and $\lambda_2 = \lambda_3 = 0$

The shear stiffness component $\mathbf{C}'_{ss}$ can be obtained from a direct shear test. Desai et al. assumed that $\mathbf{C}^i_{ss}$ is composed of a shear modulus G^i with the following definition:

$$G'(\sigma_n, \tau, u_\tau) = \frac{\partial[\tau(\sigma_r, u_r)]}{\partial u_r} \times t \Big|_{\sigma_n} \tag{4.53}$$

where: t = the thickness of the element
u_r = a relative displacement

One advantage of a thin-layer element is that it can be formulated by assuming it to be not only linear elastic or non-linear elastic but also elasto-plastic. The procedure to develop an element stiffness matrix is similar to that for a standard solid element. If $\mathbf{k}^i$ denotes the stiffness matrix, it can be expressed by:

$$\mathbf{k}' = \int_{vol} \mathbf{B}^T (\mathbf{C}^{ep})' dV \tag{4.54}$$

where: $\mathbf{B}$ = a transformation matrix
$(\mathbf{C}^{ep})'$ = the constitutive matrix for an interface element

The element equilibrium equation is then written as:

$$\mathbf{k}'\mathbf{u} = \mathbf{r} \tag{4.55}$$

where: $\mathbf{u}$ = column matrices of element displacements
$\mathbf{r}$ = column matrices of nodal forces

In linear elastic cases, $(\mathbf{C}^{ep})'$ in equation (4.54) can be simplified as:

$$(\mathbf{C}^{ep})' = (\mathbf{C}^e)' = \begin{bmatrix} C_1 & C_2 & C_2 & 0 & 0 & 0 \\ C_2 & C_1 & C_2 & 0 & 0 & 0 \\ C_2 & C_2 & C_1 & 0 & 0 & 0 \\ 0 & 0 & 0 & G_1' & 0 & 0 \\ 0 & 0 & 0 & 0 & G_2' & 0 \\ 0 & 0 & 0 & 0 & 0 & G_3' \end{bmatrix} = \begin{bmatrix} \mathbf{C}_{nn}' & 0 \\ 0 & \mathbf{C}_{ss}' \end{bmatrix} \tag{4.56}$$

where:

$$C_1 = \frac{E(1-v)}{(1+v)(1-2v)}$$

$$C_2 = \frac{Ev}{(1+v)(1-2v)}$$

E = elastic modulus

v = Poisson's ratio

G_j' ($j = 1,2,3$) = shear moduli defined by equation (4.53)

In most cases, the shear behavior is assumed to be isotropic, i.e., $G_1' = G_2' = G_3'$. As to two-dimensional plane strain idealization, equation (4.56) shrinks to:

$$(\mathbf{C}^{ep})' = (\mathbf{C}^e)' = \begin{bmatrix} C_1 & C_2 & 0 \\ C_2 & C_1 & 0 \\ 0 & 0 & G' \end{bmatrix} \tag{4.57}$$

In non-linear cases, E, v and G may be defined as a function of stress and/or strain levels. The form of the function depends upon which constitutive model is chosen for representing the material properties. For details, refer to Section 4.4.

In elasto-plastic cases, the constitutive matrix for the interface may be expressed as:

$$(\mathbf{C}^{ep})' = (\mathbf{C}^e(\mathbf{k}_s, \mathbf{k}_n))' - (\mathbf{C}^p(\mathbf{k}_s, \mathbf{k}_n, d\mathbf{u}_r^p))' \tag{4.58}$$

where: $d\mathbf{u}_r^p$ = column matrix of incremental relative displacements

The second term on the right-hand side of above equation depends upon the yield and flow criteria of soil at the interface. Mohr-Coulomb criterion can be used as yield function f as well as plastic potential function g for associative plasticity.

4.4 SHEAR STRESS-DISPLACEMENT RELATION FOR AN INTERFACE ELEMENT

In all types of interface elements introduced in the preceding sections, a constitutive relation actually means the relation between shear stress and displacement. This relation is essential in determining the shear modulus G^i, as shown in equation (4.53). Here, some typical models are introduced.

• Desai-Siriwardane Model
Desai and Siriwardane (1984) developed an elasto-perfect plastic interface model given by the following equation:

$$\tau = \begin{cases} Ks & s \leq s_0 \\ \mu\sigma_n & s > s_0 \end{cases} \tag{4.59}$$

where: τ = shear stress
σ = normal stress
s = relative displacement
μ = friction coefficient
K = elastic modulus.

In this model, the shear stress initially increases linearly with the relative displacement before the shear stress reaches its maximum value. After that, perfect plasticity occurs, that is, the shear stress remains unchanged as the relative displacement increases, as shown in Fig. 4.10. The tangent shear modulus is assumed to be zero after the maximum shear stress is reached.

The primary weakness of this model is its simplicity without considering non-linearity at an interface. In addition, a zero modulus might cause some inconveniences in a finite element analysis.

• Bekker Model
Bekker (1960) proposed an exponential stress-displacement relation for loose soil expressed by:

$$\tau = \tau_{max}\left(1 - e^{-\frac{s}{K}}\right) \tag{4.60}$$

where: τ_{max} = maximum shear stress
K = displacement modulus

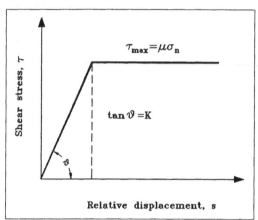

Figure 4.10 Shear stress versus relative displacement in the Desai-Siriwardane model.

This model represents a non-linear relation between shear stress and relative displacement at a soil-metal interface, as illustrated in Fig. 4.11. Zhang et al. (1986) modified this model to describe the friction between wheat kernel and its bin wall. The modified model is of the form:

$$\tau = \tau_{max}\left(1 - e^{-\sqrt{s/K}}\right) \tag{4.61}$$

♦ Extended Clough-Duncan Model

One of the basic formula in the Clough-Duncan model (1971), which is similar to the hyperbolic stress-strain relation used in the Duncan-Chang model (1970), is of the form:

$$\tau = \frac{s}{a + b\,s} \tag{4.62}$$

where: τ = interface shear stress
s = interface relative displacement
a and b = empirical coefficients determined from laboratory tests

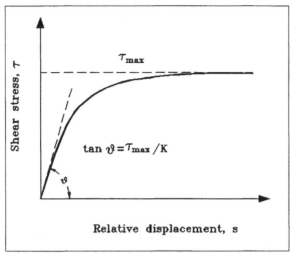

Figure 4.11 Shear stress versus relative displacement in the Bekker model.

Fig. 4.12 schematically shows the relation between τ and s represented by the above equation.

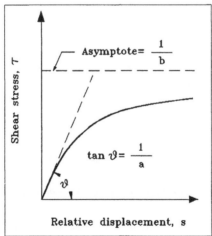

Figure 4.12 Shear stress versus relative displacement in the Clough-Duncan model.

Two parameters a and b in equation (4.62) define the shape of the stress-displacement curve. The equation (4.62) can be transformed into the following linear form:

$$y = a + b\,x$$
$$y = \frac{s}{\tau}; \quad x = s \tag{4.63}$$

from which a and b are easily determined on the basis of experimental data, as shown in Fig. 4.13. According to the role that a and b play in defining the stress-displacement curve, they are defined as:

$$a = \frac{1}{E_{it}}; \quad b = \frac{1}{\tau_{ult}} = \frac{R_{if}}{\tau_f} \tag{4.64}$$

where: E_{ii} = initial tangent shear modulus
τ_f ant τ_{ult} = measured shear stress at failure and its asymptote, respectively
R_{if} = failure ratio for an interface element

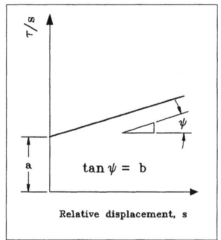

Figure 4.13 Transformed relation of shear stress versus relative displacement.

The experimental studies by Janbu (1963) indicate that E_{ii} can be expressed as:

$$E_{ii} = K_i \, \gamma_w \left(\frac{\sigma_3}{p_a} \right)^{n_i} \tag{4.65}$$

where: K_i and n_i = dimensionless stiffness number and stiffness exponent, respectively

γ_w = unit weight of water expressed in the same unit as E_{ii}

σ_3 and p_a = minor principal stress and atmospheric pressure, respectively

The shear strength at the interface is expressed as:

$$\tau_f = c_a + \sigma_n \tan \delta \qquad (4.66)$$

where: c_a and δ = adhesion and external friction angle at an interface, separately

σ_n and τ_f = normal stress and shear stress at failure, respectively

By substituting equations (4.63)-(4.66) into equation (4.62), the shear stress-displacement relation in the extended Clough-Duncan model is obtained as follows:

$$\tau = \frac{s}{\left[K_i^{-1} \gamma_w^{-1} \left(\dfrac{\sigma_n}{p_a} \right)^{-n_i} + \dfrac{s \, R_{if}}{c_a + \sigma_n \tan \delta} \right]} \qquad (4.67)$$

For incremental stress analyses, the tangent modulus, E_i, is expressed as

$$E_i = \frac{\partial \tau}{\partial s} \qquad (4.68)$$

By substituting equation (4.67) into equation (4.68) and several derivations, Equation (4.68) can finally be written as:

$$E_i = \left[1 - \frac{R_{if} \tau}{c_a + \sigma_n \tan \delta} \right]^2 K_i \gamma_w \left(\frac{\sigma_n}{p_a} \right)^{n_i} \qquad (4.69)$$

Equations (4.67) and (4.69) are the ordinary and incremental form of the extended Clough-Duncan model, respectively. Researchers usually adopted the incremental form in non-linear analyses, because of its effectiveness.

4.5   CASE STUDY: NUMERICAL SIMULATION OF FRICTION BEHAVIOR AT A SOIL-TOOL INTERFACE

Friction is one of serious causes of energy losses in agricultural production. A strategy for tribology in Canada stated that the potential saving resulting from reducing friction in agricultural operation would account to $104 million per year (NRCC, 1986). Tillage is a typical agricultural operation in which friction occurs at soil-tool interface. The objective of this section is to analyze stress distribution at soil-tool interface using finite element method and to establish a method to evaluate maximum potential saving in tool draft by reducing friction at soil-tool interface.

4.5.1 Finite Element Model

Four-node isoparametric elements are used to simulate soil dynamic response. The constitutive relation of the soil element is described using the rate-dependent hyperbolic model in Section 3.2.4. The soil-tool interface elements are implemented using the thin-layer element in Section 4.3. The constitutive relation for this interface element is implemented using a modified rate-dependent model similar to the Clough-Duncan model in Section 4.4. The finite element mesh is shown in Fig. 4.14. Soil parameters for constitutive models are listed in Table 4.1. The finite element algorithm in Section 5.4 is used for analyzing soil dynamic response.

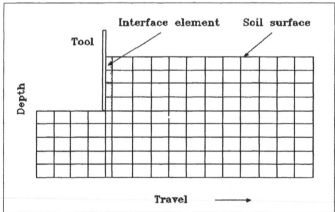

Figure 4.14   FEM mesh.

Table 4.1 Soil parameters for constitutive models

Soil parameters:	
K_s	21.2
n_s	0.0
B_{sf}	0.093
cohesion (kPa)	6.0
int. friction angle (°)	49.0
bulk density (kg/m^3)	1070
Poisson's ratio	0.32
failure ratio	0.8
Soil-metal parameters:	
K_i	2.65
n_i	0.84
B_{if}	0.093
adhesion (kPa)	3.0
ext. friction angle (°)	23.5
failure ratio	0.89

4.5.2 Analytical Results and Discussion

Normal Force Distribution
During the finite element analysis, the node forces at the soil-tool interface are calculated at each displacement increment. Fig. 4.15a shows the distribution of normal node forces at the soil-tool interface at a very low travel speed, 0.003 m/s. It can be seen that at this travel speed the normal force gradually increases from the top to the bottom of the tool, and the normal force at the bottom is always largest for all displacement steps. At the first displacement step, only the forces at the bottom area of the tool are noticeable. With the tool moving further, the forces at the whole interface increase and become significant. At the final displacement step, the increase in the forces tend to stop and reach the failure limit.

For the speeds, 1 m/s and 5 m/s, as shown in Figs. 4.15b and 4.15c, the normal force distribution pattern is different from that at 0.001 m/s. In these two cases, the forces in the middle area of the soil-tool interface become the largest compared to the forces at top and bottom area of the tool. The main reason leading to this kind of distribution is the contribution of the considerable soil acceleration caused by the rapid tool movement.

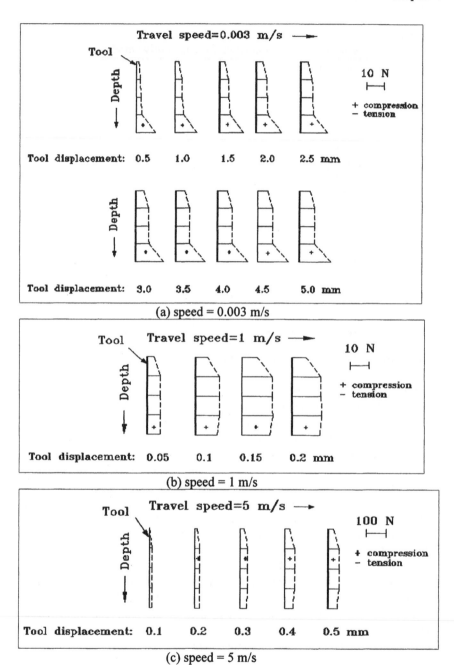

Figure 4.15 Distribution of normal forces at soil-tool interface.

Tangential Force Distribution
The tangential force distribution at speed 0.001 m/s is shown in Fig. 4.16a. The distribution pattern varies with the displacement step. At the first two steps, downward tangential forces exist at the bottom of the tool. The soil tensile strength between elements near and beneath the bottom may be the main cause of the downward tangential force. For the following three steps, all tangential forces are upward and the force gradually increases from the bottom to the top of the tool. This is due mainly to the fact that the tangential displacement at the top of the tool is larger than that at the bottom of the tool. During the rest of the steps, the distribution pattern is changed to another form. The forces in the middle area of the tool become least because of the large normal stress at the bottom and large tangential displacement at the top of the tool.

At higher speeds, 1 m/s and 5 m/s, the force distributions are indicated in Figs. 4.16b and 4.16c. In the top area of the tool, an upward tangential force exists due to the upward tangential displacement; While at the bottom, downward tangential force occurs which may be caused by the soil tensile strength between elements near the bottom of the tool. In the middle area of the tool, the forces are almost equal to zero.

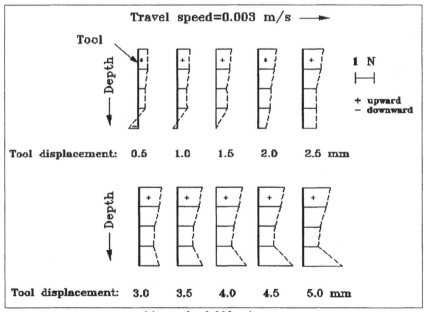

(a) speed = 0.003 m/s

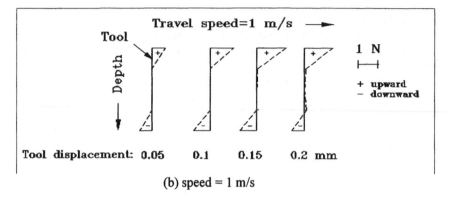

(b) speed = 1 m/s

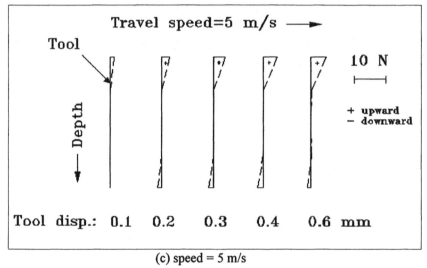

(c) speed = 5 m/s

Figure 4.16 Distribution of tangential forces at soil-tool interface.

Tangential Displacement Distribution
The tangential displacement gradually increases from the bottom to the top of the tool for all displacement steps at 0.001 m/s, as shown in Fig. 4.17a. With the travel speed increasing, the distribution pattern is changed. Figs. 4.17b and 4.17c indicate that an upward tangential displacement occurs only in the top area of the tool.

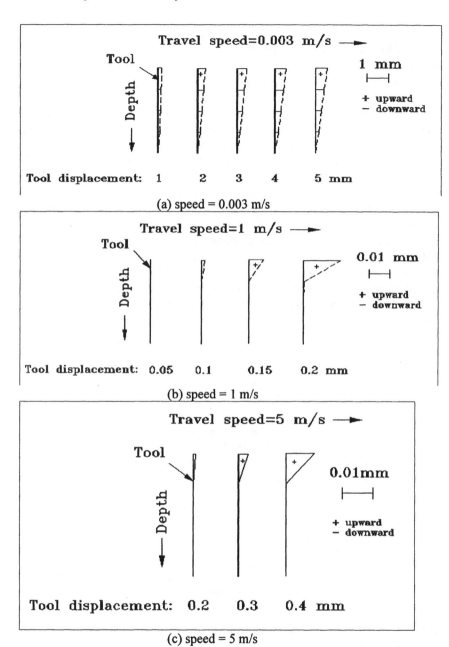

Figure 4.17 Distribution of tangential displacements at soil-tool interface.

Effect of Friction on the Tool Draft

In soil tillage operation, friction always exists at soil-tool interface. Researchers would like to reduce or eliminate this friction in order to decrease tool draft. However, by experimental methods, it is difficult to estimate the extent of maximum potential saving in tool draft for different tools and soil conditions. Using the finite element method, the simulation is carried out for two working conditions. One condition reflects the real frictional behavior between soil and tool by using the interface element. The other is designed for an absolutely smooth soil-tool interface without any friction. This can be achieved by freeing the vertical constraints of nodes at the interface in the FEM analysis. The results under quasi-static state are shown in Fig. 4.18 in which the maximum potential saving in tool draft was 24.5%. Fig. 4.19 shows the normal and tangential forces on the soil-tool interface at different speeds in cases with and without friction. The percentage of maximum potential saving in tool draft decreases with speed. The reason for this phenomenon is due mainly to the fact that at high speeds, the inertial force becomes a main contribution to the total tool draft and is not highly relevant to the condition of soil-tool interface.

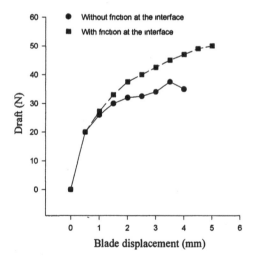

Figure 4.18 Draft vs. tool displacement with and without friction under quasi-static state.

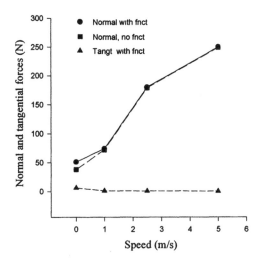

Figure 4.19 Normal and tangential forces vs. speed with and without friction.

5

ALGORITHMS IN STATIC, DYNAMIC, LINEAR, AND NON-LINEAR
FINITE ELEMENT ANALYSES

The finite element analysis of an engineering problem is usually a complex process. The most straightforward, or brute-force, approach is to try to solve the problem within a single phase. It would be difficult to use single phase because the solving process involves too many things from methodology design to implementation details. A better alternative is to use the divide-and-conquer approach to divide an entire problem-solving process into a methodology design phase and an implementation phase. During the methodology design phase, we neglect the implementation details of computer languages and physical problems. Then we concentrate on the design of methods for solving problems that are suited for computer implementation, i.e., the design of algorithms.

Algorithms are composed of a finite number of steps, each of which may require one or more operations. Their characteristics include: 1) definite; 2) effective and 3) terminating. The first characteristic means that there must be no ambiguity in the algorithm we design. The second implies that a good algorithm should be effective in execution time, storage space, or both. The last represents that the running of any algorithm should be terminated after a finite period of execution time.

To express algorithms, there are four basic ways as follows:

a) Narrative description

An algorithm is simply outlined orally.

b) Flowchart

An algorithm is outlined in a pictorial format.

c) Programming language

An algorithm is expressed in a programming language such as Pascal or C. The algorithm in such a form is actually capable of being processed by a compiler. The first programming language used as an acceptable means of communicating algorithms was ALGOL 60 (Sebestra, 1989). Since the early 1960s, it remained the sole language for publishing algorithms for over 20 years.

d) Algorithmic notation

An algorithm is defined by an algorithmic language. Horowitz and Sahani (1978) introduced one type of algorithmic language, SPARK, which includes all basic data structures and statements such as assignment, condition control and loop control. It is close to Pascal or C, and an algorithm written in SPARC can be transformed into one written in an programming language through a translator. The purpose of developing SPARC is for the situation where some persons dislike a particular programming language, while some others disfavor another programming language.

Tremblay and Bunt (1979) viewed an algorithmic language as a combination of the expressiveness and intuitiveness of natural language with the logical precision of schemes such as the flowchart. They used the pseudocode, a mixture of prose and Pascal code, to describe an algorithm.

Considering most people in the field of engineering are familiar with Fortran, a mixture of prose and Fortran keywords for condition and loop controls, is used to express algorithms in this book. All Fortran keywords will appear in boldface, and a number followed by a colon at the beginning of each step is used as a label for that step.

5.1 LINEAR STATIC CASES

Linear static cases mean the situations where all external loads to a finite element system are time-independent and material properties of each element do not change during an entire loading process. The equilibrium equation in these cases can be written as follows:

$$\mathbf{K}\,\mathbf{U} = \mathbf{R} \tag{5.1}$$

where: $\mathbf{K}$ = stiffness matrix

$\mathbf{U}$ = displacement column matrix

$\mathbf{R}$ = load column matrix of the finite element system

In linear static cases, $\mathbf{U}$ and $\mathbf{R}$ are assumed to be time-independent and $\mathbf{K}$ is a constant matrix.

Basically there are two categories of algorithms for solving equation (5.1): direct solution and iterative solution. In a direct solution, equation (5.1) is solved in a predetermined manner, whereas in an iterative solution an iterative scheme is used with an initial approximation. Either solution has some advantages and disadvantages; however, the former is currently more popular and effective than the latter.

5.1.1 Direct Solution

By assuming **K**, **U** and **R** are of the following forms:

$$
\mathbf{K} = \begin{bmatrix} k_{11} & k_{12} & \cdots & k_{1n} \\ k_{21} & k_{22} & \cdots & k_{2n} \\ \vdots & \vdots & & \\ k_{n1} & k_{n2} & \cdots & k_{nn} \end{bmatrix}, \quad \mathbf{U} = \begin{bmatrix} u_1 \\ u_2 \\ \vdots \\ u_n \end{bmatrix}, \quad \mathbf{R} = \begin{bmatrix} r_1 \\ r_2 \\ \vdots \\ r_n \end{bmatrix}, \tag{5.2}
$$

equation (5.1) is equivalent to a set of simultaneous linear equations as follows:

$$
\begin{aligned}
k_{11}u_1 + k_{12}u_2 + \cdots + k_{1n}u_n &= r_1 \\
k_{21}u_1 + k_{22}u_2 + \cdots + k_{2n}u_n &= r_2 \\
&\vdots \\
k_{n1}u_1 + k_{n2}u_2 + \cdots + k_{nn}u_n &= r_n
\end{aligned} \tag{5.3}
$$

5.1.1.1 Gaussian Elimination

Gaussian elimination is a direct method for solving equation (5.3). The process of Gaussian elimination contains two basic phases:

a) Forward elimination

Starting from the first row ($i = 1$) of the augmented matrix $\begin{bmatrix} \mathbf{K} \vdots \mathbf{U} \end{bmatrix}$, successively divide the i-th row by its diagonal element k_{ii}, called the pivot, leaving a 1 on the diagonal, and then subtract a multiple k_{ki} of the i-th row from each k-th row ($i < k \leq n$), leaving zeros in the i-th column below the diagonal 1.

b) Back substitution

Starting from the bottom row ($i = n$) of the augmented matrix $\begin{bmatrix} \mathbf{K} \vdots \mathbf{U} \end{bmatrix}$ whose content has been modified in the forward elimination, successively subtract a multiple k_{ki} of the i-th row from each k-th row ($1 \leq k < i$), leaving zeros in the k-th column above the diagonal 1.

After these two phases, the **K** submatrix in the augmented matrix turns to an identity matrix and the solution is located in the position of the **U** submatrix.

One simple example to illustrate the process of Gaussian elimination is given below. Matrices **K** and **R** have the following value:

$$\mathbf{K} = \begin{bmatrix} 2 & 3 \\ 3 & -1 \end{bmatrix}, \quad \mathbf{U} = \begin{bmatrix} 2 \\ -8 \end{bmatrix} \tag{5.4a}$$

An augmented matrix $\begin{bmatrix} \mathbf{K} \vdots \mathbf{U} \end{bmatrix}$ is formed by merging them:

$$\begin{bmatrix} \mathbf{K} \vdots \mathbf{U} \end{bmatrix} = \begin{bmatrix} 2 & 3 & \vdots & 2 \\ 3 & -1 & \vdots & -8 \end{bmatrix} \tag{5.4b}$$

Forward elimination includes three steps:

$$\text{row } 1 \div 2 \qquad \rightarrow \begin{bmatrix} 1 & 3/2 & \vdots & 1 \\ 3 & -1 & \vdots & -8 \end{bmatrix} \qquad \text{pivot} = k_{11}$$

$$\text{row } 2 - 3 \times \text{row } 1 \qquad \rightarrow \begin{bmatrix} 1 & 3/2 & \vdots & 1 \\ 0 & -11/2 & \vdots & -11 \end{bmatrix}$$

$$\text{row } 2 \div (-11/2) \qquad \rightarrow \begin{bmatrix} 1 & 3/2 & \vdots & 1 \\ 0 & 1 & \vdots & 2 \end{bmatrix} \qquad \text{pivot} = k_{22}$$

And backward substitution involves one step:

$$\text{row } 1 - 3/2 \times \text{row } 2 \qquad \rightarrow \begin{bmatrix} 1 & 0 & \vdots & -2 \\ 0 & 1 & \vdots & 2 \end{bmatrix}$$

The solution is in the third column, i.e.,

$$U = \begin{bmatrix} u_1 \\ u_2 \end{bmatrix} = \begin{bmatrix} -2 \\ 2 \end{bmatrix} \tag{5.4c}$$

The above process can be formalized by the algorithm Gaussian Elimination.

Algorithm 5.1: Gaussan Elimination

Given n, $m = n + 1$, and an augmented array of the form:

$$[K \vdots R] = \begin{bmatrix} k_{11} & k_{12} & \cdots & k_{1n} & \vdots & r_1 \\ k_{21} & k_{22} & \cdots & k_{2n} & \vdots & r_2 \\ \vdots & \vdots & & \vdots & \vdots & \vdots \\ k_{n1} & k_{n2} & \cdots & k_{nn} & \vdots & r_n \end{bmatrix} = \begin{bmatrix} k_{11} & k_{12} & \cdots & k_{1n} & \vdots & k_{1m} \\ k_{21} & k_{22} & \cdots & k_{2n} & \vdots & k_{2m} \\ \vdots & \vdots & & \vdots & \vdots & \vdots \\ k_{n1} & k_{n2} & \cdots & k_{nn} & \vdots & k_{nm} \end{bmatrix}, \quad (5.5)$$

this algorithm will produce the corresponding solution matrix:

$$\begin{bmatrix} 1 & 0 & \cdots & 0 & \vdots & u_1 \\ 0 & 1 & \cdots & 0 & \vdots & u_2 \\ \vdots & \vdots & & \vdots & \vdots & \vdots \\ 0 & 0 & \cdots & 1 & \vdots & u_n \end{bmatrix} \quad (5.6)$$

where: U = solution matrix of $KU = R$

i and k = subscripts to the rows of the augmented matrix

j = subscript to the column, as illustrated in Fig. 5.1

```
C.........Forward elimination
1:        DO 6 i = 1, n
c..........divide each element in the i-th row by the pivot, k_ii
2:        DO 3 j = i, m
3:           k_ij  ←  k_ij / k_ii
c..........subtract a multiple k_ki of the i-th row from each k-th row ( i < k ≤ n )
4:        DO 6 k = i + 1, n
5:           DO 6 j = i, m
6:              k_kj  ←  k_kj - k_ki × k_ij
C.........Back substitution
7:        DO 10 i = n, 2, -1
8:           DO 10 k = 1, i-1
9:              k_km  ←  k_km - k_ki × k_im
10:          k_ki  ←  0
11:       END
```

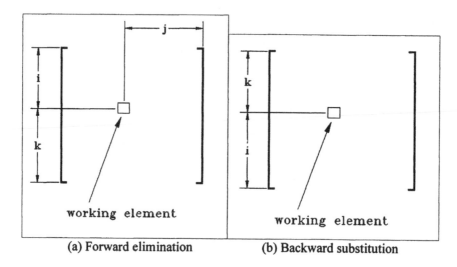

(a) Forward elimination   (b) Backward substitution

Figure 5.1 Activity scope of variable i, j and k in the augmented matrix.

The prerequisite of the Gaussian elimination is that matrix **K** is positive definite, such that the submatrix **K** in the augmented matrix $[\mathbf{K} \vdots \mathbf{U}]$ will be transformed into an identity submatrix and equation (5.3) will be solvable. Modifying matrix **K** properly with the constrain conditions in a finite element system would produce a positive definite matrix. In this chapter, it is assumed that the matrix **K** is already positive definite. With a positive definite and symmetric matrix **K**, the Gaussian elimination is numerically stable. However, if the diagonal element is not the largest one in absolute value in each row of matrix **K**, more roundoff errors may be introduced than in the case where the diagonal element is the largest.

5.1.1.2 Gaussian Elimination with Pivoting
The round off errors mentioned above can be reduced by the implementation of either partial or fully pivoting, that is, by interchanging either only rows or rows and columns, of matrix **K**, such that the next pivot is largest in absolute value. An interchange of rows does not influence the order of the solution submatrix **U** in the augmented matrix, while an interchange of columns does. If the i-th column is exchanged with the j-th column, u_i and u_j in the solution submatrix should swift their positions.

A set of three-variable equations is given below to demonstrate the process of partial and full pivoting. As a reference, the process without pivoting is also provided. To compare the round off errors of different schemes, it is assumed that for a floating point number, only four digits to the right of its decimal point are allowed to be stored in a memory cell, and the remaining digits are rounded to the fourth digit to the right of the decimal point.

$$u_1 + 3u_2 - 2u_3 = 5$$
$$2u_1 - u_2 + 5u_3 = 6 \qquad\qquad (5.7)$$
$$-3u_1 + 2u_2 + u_3 = 5$$

Without pivoting:

$$\begin{bmatrix} 1 & 3 & -2 & \vdots & 5 \\ 2 & -1 & 5 & \vdots & 6 \\ -3 & 2 & 1 & \vdots & 5 \end{bmatrix}$$

row2 + (- 2) × row1 →
row3 + (+3)× row1   →

$$\begin{bmatrix} 1 & 3 & -2 & \vdots & 5 \\ 0 & -7 & 9 & \vdots & -4 \\ 0 & 11 & -5 & \vdots & 20 \end{bmatrix}$$

row2 / (-7) →

$$\begin{bmatrix} 1 & 3 & -2 & \vdots & 5 \\ 0 & 1 & -1.2857 & \vdots & 0.5714 \\ 0 & 11 & -5 & \vdots & 20 \end{bmatrix}$$

row3 + (-11) × row2 →

$$\begin{bmatrix} 1 & 3 & -2 & \vdots & 5 \\ 0 & 1 & -1.2857 & \vdots & 0.5714 \\ 0 & 0 & 9.1427 & \vdots & 13.7146 \end{bmatrix}$$

row3 / 9.1427 →

$$\begin{bmatrix} 1 & 3 & -2 & \vdots & 5 \\ 0 & 1 & -1.2857 & \vdots & 0.5714 \\ 0 & 0 & 1 & \vdots & 1.5001 \end{bmatrix}$$

row1 + 2 × row3   →
row2 + 1.2857 × row3 →

$$\begin{bmatrix} 1 & 3 & 0 & \vdots & 8.0002 \\ 0 & 1 & 0 & \vdots & 2.5001 \\ 0 & 0 & 1 & \vdots & 1.5001 \end{bmatrix}$$

row1 + (-3) × row2   →

$$\begin{bmatrix} 1 & 0 & 0 & \vdots & 0.4999 \\ 0 & 1 & 0 & \vdots & 2.5001 \\ 0 & 0 & 1 & \vdots & 1.5001 \end{bmatrix}$$

With partial pivoting:

$$\begin{bmatrix} 1 & 3 & -2 & \vdots & 5 \\ 2 & -1 & 5 & \vdots & 6 \\ -3 & 2 & 1 & \vdots & 5 \end{bmatrix}$$

row1 ↔ row3   →

$$\begin{bmatrix} -3 & 2 & 1 & \vdots & 5 \\ 2 & -1 & 5 & \vdots & 6 \\ 1 & 3 & -2 & \vdots & 5 \end{bmatrix}$$

row1 / (-3)   →

$$\begin{bmatrix} 1 & -0.6667 & -0.3333 & \vdots & -1.6667 \\ 2 & -1 & 5 & \vdots & 6 \\ -3 & 2 & 1 & \vdots & 5 \end{bmatrix}$$

row2 + (-2) × row1   →
row3 + (-1) × row1   →

$$\begin{bmatrix} 1 & -0.6667 & -0.3333 & \vdots & -1.6667 \\ 0 & 0.3334 & 5.6666 & \vdots & 9.3334 \\ 0 & 3.6667 & -1.6667 & \vdots & 6.6667 \end{bmatrix}$$

row2 ↔ row3   →

$$\begin{bmatrix} 1 & -0.6667 & -0.3333 & \vdots & -1.6667 \\ 0 & 3.6667 & -1.6667 & \vdots & 6.6667 \\ 0 & 0.3334 & 5.6666 & \vdots & 9.3334 \end{bmatrix}$$

row2 / 3.6667   →

$$\begin{bmatrix} 1 & -0.6667 & -0.3333 & \vdots & -1.6667 \\ 0 & 1 & -0.4546 & \vdots & 1.8182 \\ 0 & 0.3334 & 5.6666 & \vdots & 9.3334 \end{bmatrix}$$

row3 + (-0.3334) × row2   →

$$\begin{bmatrix} 1 & -0.6667 & -0.3333 & \vdots & -1.6667 \\ 0 & 1 & -0.4546 & \vdots & 1.8182 \\ 0 & 0 & 5.8182 & \vdots & 8.7272 \end{bmatrix}$$

row3 / 5.8182   →

$$\begin{bmatrix} 1 & -0.6667 & -0.3333 & \vdots & -1.6667 \\ 0 & 1 & -0.4546 & \vdots & 1.8182 \\ 0 & 0 & 1 & \vdots & 1.5000 \end{bmatrix}$$

row1 + (+0.3333) × row3 →
row2 + (+0.4546) × row3 →

$$\begin{bmatrix} 1 & -0.6667 & 0 & \vdots & -1.1668 \\ 0 & 1 & 0 & \vdots & 2.5001 \\ 0 & 0 & 1 & \vdots & 1.5000 \end{bmatrix}$$

row1 + (+0.6667) × row3 →

$$\begin{bmatrix} 1 & 0 & 0 & \vdots & 0.5000 \\ 0 & 1 & 0 & \vdots & 2.5001 \\ 0 & 0 & 1 & \vdots & 1.5000 \end{bmatrix}$$

With full pivoting:

$$\begin{bmatrix} 1 & 3 & -2 & \vdots & 5 \\ 2 & -1 & 5 & \vdots & 6 \\ -3 & 2 & 1 & \vdots & 5 \end{bmatrix}$$

row1 ↔ row3 →

$$\begin{bmatrix} -3 & 2 & 1 & \vdots & 5 \\ 2 & -1 & 5 & \vdots & 6 \\ 1 & 3 & -2 & \vdots & 5 \end{bmatrix}$$

row1 / (-3) →

$$\begin{bmatrix} 1 & -0.6667 & -0.3333 & \vdots & -1.6667 \\ 2 & -1 & 5 & \vdots & 6 \\ -3 & 2 & 1 & \vdots & 5 \end{bmatrix}$$

row2 + (-2) × row1 →
row3 + (-1) × row1 →

$$\begin{bmatrix} 1 & -0.6667 & -0.3333 & \vdots & -1.6667 \\ 0 & 0.3334 & 5.6666 & \vdots & 9.3334 \\ 0 & 3.6667 & -1.6667 & \vdots & 6.6667 \end{bmatrix}$$

column1 ↔ column3 →

$$\begin{bmatrix} 1 & -0.3333 & -0.6667 & \vdots & -1.6667 \\ 0 & 5.6666 & 0.3334 & \vdots & 9.3334 \\ 0 & -1.6667 & 3.6667 & \vdots & 6.6667 \end{bmatrix}$$

row2 / 5.6666 →

$$\begin{bmatrix} 1 & -0.3333 & -0.6667 & \vdots & -1.6667 \\ 0 & 1 & 0.0588 & \vdots & 1.6471 \\ 0 & -1.6667 & 3.6667 & \vdots & 6.6667 \end{bmatrix}$$

row3 + (+1.6667) × row2 →
$$\begin{bmatrix} 1 & -0.3333 & -0.6667 & \vdots & -1.6667 \\ 0 & 1 & 0.0588 & \vdots & 1.6471 \\ 0 & 0 & 3.7647 & \vdots & 9.4119 \end{bmatrix}$$

row3 / 3.7647 →
$$\begin{bmatrix} 1 & -0.3333 & -0.6667 & \vdots & -1.6667 \\ 0 & 1 & 0.0588 & \vdots & 1.6471 \\ 0 & 0 & 1 & \vdots & 2.5001 \end{bmatrix}$$

row1 + (+0.6667) × row3 →
row2 + (-0.0588) × row3 →
$$\begin{bmatrix} 1 & -0.3333 & 0 & \vdots & 0.0001 \\ 0 & 1 & 0 & \vdots & 1.5001 \\ 0 & 0 & 1 & \vdots & 2.5001 \end{bmatrix}$$

row1 + (+0.3333) × row2 →
$$\begin{bmatrix} 1 & 0 & 0 & \vdots & 0.5001 \\ 0 & 1 & 0 & \vdots & 1.5001 \\ 0 & 0 & 1 & \vdots & 2.5001 \end{bmatrix}$$

Since column 2 is exchanged with column 3 in full pivoting scheme, u_2 and u_3 swift their positions in the solution submatrix as illustrated above. The exact solution for Equation (5.7) is $u_1 = 0.5$, $u_2 = 2.5$ and $u_3 = 1.5$. The total absolute errors of u_1, u_2 and u_3 for the Gaussian elimination without pivoting, with partial pivoting and with full pivoting, are 0.0003, 0.0001 and 0.0003, respectively. Pivoting may reduce the round off error, but not always.

The reason for advocating the pivoting is that severe computational errors can arise if the k_{ii} in Algorithm 5.1, which is used to scale a row, is very small in absolute value. If k_{ii} is very small, the scaling factor k_{ij} / k_{ii} at label 3 in Algorithm 5.1, which is used to eliminate a multiple of the elements in the i-th rows from the elements in the k-th rows (h), will be very large. In fact, it could be so large as to dwarf the actual coefficients k_{kj} on the right side of the assignment statement at label 6 in Algorithm 5.1, to the point where the new value of k_{kj} on the left side gets distorted by round off errors.

The algorithm of the Gaussian elimination with partial pivoting is given below.

Algorithm 5.2: Gaussan Elimination with partial pivoting
The definition of variables and matrices is the same as in Algorithm 5.1. For each i-th row ($1 \le i \le n$), this algorithm scans down the i-th column to find the largest element in the k-th rows ($i \le k \le n$). The row containing the element is

exchanged with the i-th row, and then a multiple of elements in the i-th row is eliminated from the elements in the k-th rows exactly as in Algorithm 5.1. As to the backward substitution phase, no change is necessary.

C..........Forward elimination
1:          **DO** 13 $i = 1, n$
c..........find the largest element in the i-th column
2:          max $\leftarrow$ i
3:          **DO** 4 $k = i + 1, n$
4:          IF $(|k_{ki}| > |k_{maxi}|)$ max $\leftarrow k$

c..........swift the row corresponding to max with the i-th row
5:          DO 8 $j = i$, m
6:          $temp \leftarrow k_{ij}$
7:          $k_{ij} \leftarrow k_{max\,i}$
8:          $k_{max\,i} \leftarrow temp$

c..........divide each element in the i-th row by the pivot, k_{ii}
9:          DO 10 $j = i$, m
10:         $k_{ij} \leftarrow k_{ij} / k_{ii}$

c..........subtract a multiple k_{ki} of the i-th row from each k-th row ($i < k \le n$)
11:         **DO** 13 $k = i + 1$, n
12:             **DO** 13 $j = i$, m
13:             $k_{kj} \leftarrow k_{kj} - k_{ki} \times k_{ij}$
C..........Back substitution
            the same as in Algorithm 5.1

5.1.1.3 Gaussian-Jordan Elimination

Gaussian-Jordan elimination is a variation of the Gaussian scheme. Its unique feature is that the matrix elements both above and below a particular pivot are eliminated to zeros at the same time, such that a backward substitution becomes unnecessary.

Its process is illustrated below using the same set of equations as equation (5.7).

Without pivoting:

$$\begin{bmatrix} 1 & 3 & -2 & \vdots & 5 \\ 2 & -1 & 5 & \vdots & 6 \\ -3 & 2 & 1 & \vdots & 5 \end{bmatrix}$$

$$
\begin{array}{l}
\text{row2} + (-2) \times \text{row1} \quad \rightarrow \\
\text{row3} + (+3) \times \text{row1} \quad \rightarrow
\end{array}
\qquad
\begin{bmatrix}
1 & 3 & -2 & \vdots & 5 \\
0 & -7 & 9 & \vdots & -4 \\
0 & 11 & -5 & \vdots & 20
\end{bmatrix}
$$

$$
\text{row2} / (-7) \quad \rightarrow
\qquad
\begin{bmatrix}
1 & 3 & -2 & \vdots & 5 \\
0 & 1 & -1.2857 & \vdots & 0.5714 \\
0 & 11 & -5 & \vdots & 20
\end{bmatrix}
$$

$$
\begin{array}{l}
\text{row1} + (-3) \times \text{row2} \quad \rightarrow \\
\\
\text{row3} + (-11) \times \text{row2} \quad \rightarrow
\end{array}
\qquad
\begin{bmatrix}
1 & 0 & 1.8571 & \vdots & 3.2858 \\
0 & 1 & -1.2857 & \vdots & 0.5714 \\
0 & 0 & 9.1427 & \vdots & 13.7146
\end{bmatrix}
$$

$$
\text{row3} / 9.1427 \quad \rightarrow
\qquad
\begin{bmatrix}
1 & 0 & 1.8571 & \vdots & 3.2858 \\
0 & 1 & -1.2857 & \vdots & 0.5714 \\
0 & 0 & 1 & \vdots & 1.5001
\end{bmatrix}
$$

$$
\begin{array}{l}
\text{row1} + (-1.8571) \times \text{row3} \quad \rightarrow \\
\text{row2} + (+1.2857) \times \text{row3} \quad \rightarrow
\end{array}
\qquad
\begin{bmatrix}
1 & 0 & 0 & \vdots & 0.5000 \\
0 & 1 & 0 & \vdots & 2.5001 \\
0 & 0 & 1 & \vdots & 1.5001
\end{bmatrix}
$$

With partial pivoting:

$$
\begin{bmatrix}
1 & 3 & -2 & \vdots & 5 \\
2 & -1 & 5 & \vdots & 6 \\
-3 & 2 & 1 & \vdots & 5
\end{bmatrix}
$$

$$
\text{row1} \leftrightarrow \text{row3} \quad \rightarrow
\qquad
\begin{bmatrix}
-3 & 2 & 1 & \vdots & 5 \\
2 & -1 & 5 & \vdots & 6 \\
1 & 3 & -2 & \vdots & 5
\end{bmatrix}
$$

row1 / (-3) $\rightarrow$

$$\begin{bmatrix} 1 & -0.6667 & -0.3333 & \vdots & -1.6667 \\ 2 & -1 & 5 & \vdots & 6 \\ -3 & 2 & 1 & \vdots & 5 \end{bmatrix}$$

row2 + (-2) × row1 $\rightarrow$
row3 + (-1) × row1 $\rightarrow$

$$\begin{bmatrix} 1 & -0.6667 & -0.3333 & \vdots & -1.6667 \\ 0 & 0.3334 & 5.6666 & \vdots & 9.3334 \\ 0 & 3.6667 & -1.6667 & \vdots & 6.6667 \end{bmatrix}$$

row2 $\leftrightarrow$ row3 $\rightarrow$

$$\begin{bmatrix} 1 & -0.6667 & -0.3333 & \vdots & -1.6667 \\ 0 & 3.6667 & -1.6667 & \vdots & 6.6667 \\ 0 & 0.3334 & 5.6666 & \vdots & 9.3334 \end{bmatrix}$$

row2 / 3.6667 $\rightarrow$

$$\begin{bmatrix} 1 & -0.6667 & -0.3333 & \vdots & -1.6667 \\ 0 & 1 & -0.4546 & \vdots & 1.8182 \\ 0 & 0.3334 & 5.6666 & \vdots & 9.3334 \end{bmatrix}$$

row1 + (+0.6667) × row2 $\rightarrow$

row3 + (-0.3334) × row2 $\rightarrow$

$$\begin{bmatrix} 1 & 0 & -0.6364 & \vdots & -0.4545 \\ 0 & 1 & -0.4546 & \vdots & 1.8182 \\ 0 & 0 & 5.8182 & \vdots & 8.7272 \end{bmatrix}$$

row3 / 5.8182 $\rightarrow$

$$\begin{bmatrix} 1 & 0 & -0.6364 & \vdots & -0.4545 \\ 0 & 1 & -0.4546 & \vdots & 1.8182 \\ 0 & 0 & 1 & \vdots & 1.5000 \end{bmatrix}$$

row1 + (+0.6364) × row3 $\rightarrow$
row2 + (+0.4546) × row3 $\rightarrow$

$$\begin{bmatrix} 1 & 0 & 0 & \vdots & 0.5001 \\ 0 & 1 & 0 & \vdots & 2.5001 \\ 0 & 0 & 1 & \vdots & 1.5000 \end{bmatrix}$$

The above process is formalized below.
Algorithm 5.3: Gauss-Jordan Elimination without pivoting
The definition of variables and matrices is the same as in Algorithm 5.1.

C..........Forward elimination

1:     **DO** 9 $i = 1$, n
c..........divide each element in the i-th row by the pivot, k_{ii}
2:     **DO** 3 $j = i$, m
3:     $k_{ij} \leftarrow k_{ij} / k_{ii}$
c..........subtract a multiple k_{ki} of the i-th row from each k-th row ($1 \le k \le i - 1$
       and $i + 1 \le k \le n$)
4:     **DO** 6 $k = 1$, i-1
5:        **DO** 6 $j = i$, m
6:        $k_{kj} \leftarrow k_{kj} - k_{ki} \times k_{ij}$
7:     **DO** 9 $k = i + 1$, n
8:        **DO** 9 $j = i$, m
9:        $k_{kj} \leftarrow k_{kj} - k_{ki} \times k_{ij}$
C..........Back substitution
       the same as in Algorithm 5.1

Compared to Algorithm 5.1, the Gauss-Jordan elimination is simpler but less efficient due to more arithmetic operations required. Pivoting can be applied to this algorithm. Either Gaussian or Gauss-Jordan elimination can be applied to both non-symmetric and symmetric matrix **K**. In most of finite element analyses, the matrix **K** is symmetric. In the following subsection, a method particularly pertaining to a symmetric **K** will be introduced.

5.1.1.4 Triangular Decomposition

The basic procedure of the Gaussian elimination is to reduce the coefficient matrix K to an upper triangular matrix in the forward elimination phase and to calculate the unknown displacement matrix U based upon this upper triangular matrix in the backward substitution phase. In this subsection, we formalize the process of triangular decomposition of K, which is a variation of the Gaussian elimination.

Consider a equilibrium equation as in equation (5.1),

$$\mathbf{K}\,\mathbf{U} = \mathbf{R} \tag{5.1}$$

If the matrix **K** is symmetric and positive definite, equation (5.1) is solvable by the Gaussian elimination without pivoting. According to linear algebra, there are three types of elementary row operations: (1) interchange of two rows; (2) multiplication of a row by a nonzero scalar; (3) addition of a scalar multiple of one row to another row. If only types 2 and 3 are allowed in transforming **K** into an upper triangular matrix **V**,

$$\mathbf{V}\,\mathbf{U} = \mathbf{Y} \tag{5.8}$$

then the transformation from equation (5.1) to equation (5.8) is equivalent to multiplying equation (5.1) on the left of $\mathbf{K}$ and $\mathbf{R}$ by a series of elementary matrices, that is,

$$\begin{aligned}
\mathbf{V}\,\mathbf{U} &= \mathbf{P}_n \times \mathbf{P}_{n-1} \times \cdots \times \mathbf{P}_2 \times \mathbf{P}_1 \times \mathbf{K}\,\mathbf{U} \\
&= \mathbf{P}_n \times \mathbf{P}_{n-1} \times \cdots \times \mathbf{P}_2 \times \mathbf{P}_1 \times \mathbf{R} = \mathbf{Y}
\end{aligned} \tag{5.9}$$

where, each elementary matrix $\mathbf{P}_i$, $1 \leq i \leq n$, is a lower triangular matrix. Let

$$\mathbf{L}^{'} = \mathbf{P}_n \times \mathbf{P}_{n-1} \times \cdots \times \mathbf{P}_2 \times \mathbf{P}_1,$$

and then $\mathbf{L}^{'}$ must be a lower triangular matrix. equation (5.9) can be rewritten as:

$$\mathbf{L}^{'}\,\mathbf{K} = \mathbf{V}$$

If $\mathbf{L}^{'-1} = \mathbf{L}$, then

$$\mathbf{K} = \mathbf{L}\,\mathbf{V} \tag{5.10}$$

where:

$$\mathbf{L} = \begin{bmatrix} L_{11} & 0 & \cdots & 0 \\ L_{21} & L_{22} & \cdots & 0 \\ \vdots & \vdots & \ddots & \vdots \\ L_{n1} & L_{n2} & \cdots & L_{33} \end{bmatrix}, \qquad \mathbf{V} = \begin{bmatrix} V_{11} & V_{12} & \cdots & V_{1n} \\ 0 & V_{22} & \cdots & V_{2n} \\ \vdots & \vdots & \ddots & \vdots \\ 0 & 0 & \cdots & V_{nn} \end{bmatrix} \tag{5.11}$$

The step represented by equation (5.10) is called the triangular decomposition of $\mathbf{K}$. Since each of $\mathbf{L}$ and $\mathbf{V}$ contains $n(n+1)/2$ elements and $\mathbf{K}$ contains only $n \times n$ elements, the decomposition is not unique. Two well-known decomposition methods are introduced below.

$\mathbf{LDL}^T$ Decomposition

$\mathbf{LDL}^T$ decomposition is basically of two forms. In the first form, $\mathbf{L}$ is restricted as a lower triangular identity matrix whose principal diagonal elements are all unity, that is,

$$\mathbf{L}^1 = \begin{bmatrix} 1 & 0 & \cdots & 0 \\ L_{21} & 1 & \cdots & 0 \\ \vdots & \vdots & \ddots & \vdots \\ L_{n1} & L_{n2} & \cdots & 1 \end{bmatrix}, \qquad \mathbf{V}^1 = \begin{bmatrix} V_{11} & V_{12} & \cdots & V_{1n} \\ 0 & V_{22} & \cdots & V_{2n} \\ \vdots & \vdots & \ddots & \vdots \\ 0 & 0 & \cdots & V_{nn} \end{bmatrix} \tag{5.12}$$

This can be realized by applying only the type 3 of elementary row operations in transforming $\mathbf{K}$ into an upper triangular matrix $\mathbf{V}^1$. $\mathbf{V}^1$ in equation (5.12) can be represented by a product of two matrices:

$$\mathbf{V}^1 = \mathbf{D}^1 \mathbf{V}^{1'} \tag{5.13}$$

where:

$$\mathbf{D}^1 = \begin{bmatrix} V_{11} & 0 & \cdots & 0 \\ 0 & V_{22} & \cdots & 0 \\ \vdots & \vdots & \ddots & \vdots \\ 0 & 0 & \cdots & V_{nn} \end{bmatrix}, \qquad \mathbf{V}^{1'} = \begin{bmatrix} 1 & V_{12}/V_{11} & \cdots & V_{1n}/V_{11} \\ 0 & 1 & \cdots & V_{2n}/V_{22} \\ \vdots & \vdots & \ddots & \vdots \\ 0 & 0 & \cdots & 1 \end{bmatrix} \tag{5.14}$$

If $\mathbf{K}$ is a symmetric matrix, we can write

$$\mathbf{K} = \mathbf{L}^1 \mathbf{V}^1 = \mathbf{L}^1 \mathbf{D}^1 \mathbf{V}^{1'} = \mathbf{L}^1 \mathbf{D}^1 \mathbf{L}^{1^T} \tag{5.15}$$

In the second form, $\mathbf{L}^2$ has the general form in equation (5.11) and $\mathbf{V}^2$ is of the form:

$$\mathbf{V}^2 = \begin{bmatrix} 1 & V_{12} & \cdots & V_{1n} \\ 0 & 1 & \cdots & V_{2n} \\ \vdots & \vdots & \ddots & \vdots \\ 0 & 0 & \cdots & 1 \end{bmatrix} \tag{5.16}$$

This can be realized by applying both type 3 and type 2 of elementary row operations in transforming $\mathbf{K}$ into an upper triangular matrix $\mathbf{V}^2$. If $\mathbf{K}$ is a symmetric matrix, we can write

$$\mathbf{K} = \mathbf{L}^2 \mathbf{V}^2 = \mathbf{L}^2 \mathbf{D}^2 \mathbf{V}^{2'} = \mathbf{L}^2 \mathbf{D}^2 \mathbf{L}^{2^T} \tag{5.17}$$

where:

$$
\mathbf{L}^2 = \begin{bmatrix} L_{11} & 0 & \cdots & 0 \\ L_{21} & L_{22} & \cdots & 0 \\ \vdots & \vdots & \ddots & \vdots \\ L_{n1} & L_{n2} & \cdots & L_{nn} \end{bmatrix}, \quad
\mathbf{D}^2 = \begin{bmatrix} 1/L_{11} & 0 & \cdots & 0 \\ 0 & 1/L_{22} & \cdots & 0 \\ \vdots & \vdots & \ddots & \vdots \\ 0 & 0 & \cdots & 1/L_{nn} \end{bmatrix} \qquad (5.18)
$$

In the following we discuss only the second form of $\mathbf{LDL}^T$ decomposition and eliminate the superscript 2. $\mathbf{L}$ and $\mathbf{D}$ can be determined by comparing elements in $\mathbf{LDL}^T$ with elements in $\mathbf{K}$. First, according to equation (5.18), $\mathbf{LD}$ is of the form:

$$
\mathbf{LD} = \begin{bmatrix} 1 & 0 & \cdots & 0 \\ L_{21}/L_{11} & 1 & \cdots & 0 \\ \vdots & \vdots & \ddots & \vdots \\ L_{n1}/L_{11} & L_{n2}/L_{22} & \cdots & 1 \end{bmatrix} \qquad (5.19)
$$

Then, if $\mathbf{K}$ is symmetric, only lower (or upper) triangular portion of $\mathbf{LDL}^T$ is needed for the comparison,

$$
\mathbf{LDL}^T = \begin{bmatrix} 1 & 0 & \cdots & 0 \\ L_{21}/L_{11} & 1 & \cdots & 0 \\ \vdots & \vdots & \ddots & \vdots \\ L_{n1}/L_{11} & L_{n2}/L_{22} & \cdots & 1 \end{bmatrix} \begin{bmatrix} L_{11} & L_{21} & \cdots & L_{n1} \\ 0 & L_{22} & \cdots & L_{n2} \\ \vdots & \vdots & \ddots & \vdots \\ 0 & 0 & \cdots & L_{nn} \end{bmatrix}
$$

$$
= \begin{bmatrix} L_{11} & & & & \\ L_{21} & \dfrac{L_{21}^2}{L_{11}}+L_{22} & & & \\ L_{31} & \dfrac{L_{31}L_{21}}{L_{11}}+\dfrac{L_{32}L_{22}}{L_{22}} & \dfrac{L_{31}^2}{L_{11}}+\dfrac{L_{32}^2}{L_{22}}+L_{33} & & \\ \vdots & \vdots & \vdots & \ddots & \\ L_{n1} & \dfrac{L_{n1}L_{21}}{L_{11}}+\dfrac{L_{n2}L_{22}}{L_{22}} & \dfrac{L_{n1}L_{31}}{L_{11}}+\dfrac{L_{n2}L_{32}}{L_{22}}+\dfrac{L_{n3}L_{33}}{L_{33}} & \cdots & \dfrac{L_{n1}^2}{L_{11}}+\dfrac{L_{n2}^2}{L_{22}}+\cdots+L_{nn} \end{bmatrix}
$$

$$
= \begin{bmatrix}
k_{11} & & & & \\
k_{21} & k_{22} & & & \\
k_{31} & k_{32} & k_{33} & & \\
\vdots & \vdots & \vdots & \ddots & \\
k_{n1} & k_{n2} & k_{n3} & \cdots & k_{nn}
\end{bmatrix}
\tag{5.20}
$$

Comparison of elements in equation (5.20) will lead to an expression for all elements in **L**:

$$
L_{ij} = k_{ij} - \sum_{k=1}^{j-1} \frac{L_{ik} L_{jk}}{L_{kk}} \qquad (i = 1,2,\cdots,n; \quad j = 1,\cdots,i)
\tag{5.21}
$$

After the decomposition of **K** is completed, the solution to equation (5.1) can now be obtained by solving the pair of equations:

$$
\mathbf{L}^T \mathbf{U} = \mathbf{Y} \tag{5.22}
$$
$$
\mathbf{L}\,\mathbf{D}\mathbf{Y} = \mathbf{R} \tag{5.23}
$$

where, $\mathbf{Y} = \begin{bmatrix} y_1 & y_2 & \cdots & y_n \end{bmatrix}^T$. equation (5.23) is called 'forward elimination' and its solution can be written as:

$$
y_i = r_i - \sum_{j=1}^{i-1} \frac{L_{ij} y_j}{L_{ij}} \qquad (i = 1,2,\cdots,n),
\tag{5.24}
$$

while equation (5.22) is called 'backward substitution' and its solution is:

$$
u_i = \frac{y_i - \sum_{k=i+1}^{n} L_{ki} u_k}{L_{ii}} \qquad (i = n, n-1,\cdots,1)
\tag{5.25}
$$

The advantage of **LDL**^{T} decomposition is that we need to calculate only one of either **L** or **L**^{T}; this is approximately half the work in the Gaussian elimination of unsymmetrical matrices.

A simple example of using **LDL**^{T} decomposition to solve a set of three-variable symmetric equations is given below.

$$K = \begin{bmatrix} 3 & -2 & 1 \\ -2 & 3 & 2 \\ 1 & 2 & 2 \end{bmatrix}, \quad R = \begin{bmatrix} 3 \\ -3 \\ 2 \end{bmatrix}$$

Using equation (5.21) to solve L_y gives:

$$L_{11} = k_{11}$$
$$L_{21} = k_{21}, \quad L_{22} = k_{22} - L_{21}^2 / L_{11}$$
$$L_{31} = k_{31}, \quad L_{32} = k_{32} - L_{31}L_{21} / L_{11}, \quad L_{33} = k_{33} - L_{31}^2 / L_{11} - L_{32}^2 / L_{22}$$

and

$$L = \begin{bmatrix} 3 & 0 & 0 \\ -2 & 5/3 & 0 \\ 1 & 8/3 & -13/5 \end{bmatrix}$$

Forward elimination:
   Using equation (5.24) gives:

$$y_1 = r_1, \quad y_2 = r_2 - L_{21} y_1 / L_{11}$$
$$y_3 = r_3 - L_{31} y_1 / L_{11} - L_{32} y_2 / L_{22}$$

and

$$Y = \begin{bmatrix} 3 \\ -1 \\ 13/5 \end{bmatrix}$$

Backward substitution:
   Using equation (5.25) leads to:

$$u_3 = \frac{y_3}{L_{33}}, \quad u_2 = \frac{y_2 - L_{32}u_3}{L_{22}}$$

$$u_1 = \frac{y_1 - L_{21}u_2 - L_{31}u_3}{L_{11}}$$

and

$$U = \begin{bmatrix} 2 \\ 1 \\ -1 \end{bmatrix}$$

Algorithm 5.5: LDL^T decomposition
K and R are known matrices with a size of $n \times n$ and $n \times 1$, respectively. **K U=
R**, where **U** is an unknown displacement matrix with a size of $n \times 1$. **K** is at
least a symmetric matrix.

C.......... LDL^T decomposition
1:      **DO** 3 $i = 1$, n
2:      **DO** 3 $j = 1$, i

3:      $L_{ij} \leftarrow k_{ij} - \sum\limits_{k=1}^{j-1} \dfrac{L_{ik} L_{jk}}{L_{kk}}$

C..........Forward elimination
4:      **DO** 5 $i = 1$, n

5:      $y_i \leftarrow r_i - \sum\limits_{j=1}^{i-1} \dfrac{L_{ij} y_j}{L_{jj}}$

C..........Back substitution
6:      **DO** 7 $i = $ n, 1, -1

7:      $u_i \leftarrow \dfrac{y_i - \sum\limits_{k=i+1}^{n} L_{ki} u_k}{L_{ii}}$

8:      **END**

Cholesky's Decomposition
When **K** is symmetric and positive definite, equation (5.10) can be rewritten as:

$$K = L L^T \tag{5.26}$$

where:

$$\mathbf{L} = \begin{bmatrix} L_{11} & 0 & \cdots & 0 \\ L_{21} & L_{22} & \cdots & 0 \\ \vdots & \vdots & \ddots & \vdots \\ L_{n1} & L_{n2} & \cdots & L_{33} \end{bmatrix}$$

Equation (5.26) can be rewritten as:

$$\mathbf{LL}^T = \begin{bmatrix} L_{11} & 0 & \cdots & 0 \\ L_{21} & L_{22} & \cdots & 0 \\ \vdots & \vdots & \ddots & \vdots \\ L_{n1} & L_{n2} & \cdots & L_{nn} \end{bmatrix} \begin{bmatrix} L_{11} & L_{21} & \cdots & L_{n1} \\ 0 & L_{22} & \cdots & L_{n2} \\ \vdots & \vdots & \ddots & \vdots \\ 0 & 0 & \cdots & L_{nn} \end{bmatrix}$$

$$= \begin{bmatrix} L_{11}^{\,2} & & & & symmetric \\ L_{11}L_{21} & L_{21}^{\,2}+L_{22}^{\,2} & & & \\ L_{11}L_{31} & L_{31}L_{21}+L_{32}L_{22} & L_{31}^{\,2}+L_{32}^{\,2}+L_{33}^{\,3} & & \\ \vdots & \vdots & \vdots & \ddots & \\ L_{11}L_{n1} & L_{n1}L_{21}+L_{n2}L_{22} & L_{n1}L_{31}+L_{n2}L_{32}+L_{n3}L_{33} & \cdots & L_{n1}^{\,2}+L_{n2}^{\,2}+\cdots+L_{nn}^{\,2} \end{bmatrix}$$

$$(5.27)$$

Comparing elements in equation (5.27) gives:

$$L_{11} = \sqrt{k_{11}}$$

$$L_{21} = \frac{k_{21}}{L_{11}}, \quad L_{22} = \sqrt{k_{22} - L_{21}^{\,2}}$$

$$L_{31} = \frac{k_{31}}{L_{11}}, \quad L_{32} = \frac{k_{32} - L_{31}L_{21}}{L_{22}}, \quad L_{33} = \sqrt{k_{33} - L_{31}^{\,2} - L_{32}^{\,2}}$$

$$\vdots$$

$$L_{n1} = \frac{k_{n1}}{L_{11}}, \quad L_{n2} = \frac{k_{n2} - L_{n1}L_{21}}{L_{22}}, \quad L_{n3} = \frac{k_{n3} - L_{n1}L_{31} - L_{n2}L_{32}}{L_{33}},$$

$$\cdots, L_{nn} = \sqrt{k_{nn} - L_{n1}^{\,2} - \cdots - L_{n,n-1}^{\,2}} \qquad (5.28)$$

Summarizing information in equation (5.28) yields:

$$L_{ij} = \begin{cases} \dfrac{1}{L_{jj}}\left(k_{ij} - \displaystyle\sum_{k=1}^{j-1} L_{ik}L_{jk}\right) & i > j \\[2em] \sqrt{k_{ii} - \displaystyle\sum_{k=1}^{i-1} L_{ik}^{2}} & i = j \end{cases} \tag{5.29}$$

Note that square root operations are needed in the Cholesky decomposition as illustrated above. With floating point numbers, each of either the square root or division operations is considered to be about 4 times longer than a multiplication in execution time (Hennessy and Patterson, 1990). If so, the time efficiency of Cholesky's decomposition is approximately the same as that of $\mathbf{LDL}^{T}$.

The equations for the forward elimination and backward substitution are, respectively, given by:

$$\mathbf{LY} = \mathbf{R} \tag{5.30}$$
$$\mathbf{L}^{T}\mathbf{U} = \mathbf{Y} \tag{5.31}$$

The solutions to above two equations are:

$$y_{i} = \frac{1}{L_{ii}}\left(r_{i} - \sum_{j=1}^{i-1} L_{ij}y_{j}\right) \qquad (i = 1,2,\cdots,n), \tag{5.32}$$

and

$$u_{i} = \frac{1}{L_{ii}}\left(y_{i} - \sum_{k=i+1}^{n} L_{ki}u_{k}\right) \qquad (i = n, n-1,\cdots,1) \tag{5.33}$$

A simple example of using Cholesky's decomposition to solve a set of three-variable symmetric equations is given below.

$$\mathbf{K} = \begin{bmatrix} 1 & 1 & 1 \\ 1 & 2 & 2 \\ 1 & 2 & 3 \end{bmatrix}, \qquad \mathbf{R} = \begin{bmatrix} 1 \\ 2 \\ 4 \end{bmatrix}$$

Using equation (5.29) to solve L_{ij} gives:

$$L = \begin{bmatrix} 1 & 0 & 0 \\ 1 & 1 & 0 \\ 1 & 1 & 1 \end{bmatrix}$$

Forward elimination:
   Using equation (5.32) gives:

$$y_1 = r_1 / L_{11}, \quad y_2 = (r_2 - L_{21} y_1) / L_{22}$$
$$y_3 = (r_3 - L_{31} y_1 - L_{32} y_2) / L_{33}$$

and

$$Y = \begin{bmatrix} 1 \\ 1 \\ 2 \end{bmatrix}$$

Backward substitution:
   Using equation (5.33) leads to:

$$u_3 = \frac{y_3}{L_{33}}, \quad u_2 = \frac{y_2 - L_{32} u_3}{L_{22}}$$
$$u_1 = \frac{y_1 - L_{21} u_2 - L_{31} u_3}{L_{11}}$$

and

$$U = \begin{bmatrix} 0 \\ -1 \\ 2 \end{bmatrix}$$

Algorithm 5.6: Cholesky's Decomposition
K and **R** are known matrices with a size of $n \times n$ and $n \times 1$, respectively. **K U= R**, where **U** is an unknown displacement matrix with a size of $n \times 1$. **K** is required to be a symmetric and positive definite matrix.

C..........Cholesky's decomposition
1:      **DO** 4 $i = 1$, n

2:     $$L_{ii} \leftarrow \sqrt{k_{ii} - \sum_{k=1}^{i-1} L_{ik}^{2}}$$

3:     **DO 4** $j = 1, i\text{-}1$

4:     $$L_{ij} \leftarrow \frac{1}{L_{jj}}\left(k_{ij} - \sum_{k=1}^{j-1} L_{ik}L_{jk}\right)$$

C..........Forward elimination

5:     **DO 6** $i = 1, n$

6:     $$y_{i} \leftarrow \frac{1}{L_{ii}}\left(r_{i} - \sum_{j=1}^{i-1} L_{ij}y_{j}\right)$$

C..........Back substitution

7:     **DO 8** $i = n, 1, -1$

8:     $$u_{i} \leftarrow \frac{1}{L_{ii}}\left(y_{i} - \sum_{k=i+1}^{n} L_{ki}u_{k}\right)$$

9:     **END**

The Cholesky decomposition is suitable to only **K** which is symmetric and positive definite, while the **LDL**^{T} decomposition is applicable to any symmetric **K** matrix.

5.1.2 Iterative Solution

In this section, we will examine two iterative methods for solving a set of simultaneous linear equations. Compared to direct methods, the advantages of iterative methods include the easy implementation of computation and possible shorter execution time when **K** is very sparse, that is there are very few nonzero elements in **K**.

5.1.2.1 Jacobi Method
The Jacobi method includes two primary steps:
(1) Transformation of original equations
   An original set of equations has to be transformed into an iterative form such that an iteration can be proceeded. To accomplish this, one of the variables in each equation is solved. The selection criterion of a variable in each equation is to choose one with the largest coefficient in a sense of absolute value and so far not solved. For instance, the following set of equations:

$$u_1 + 3u_2 - 2u_3 = 5 \tag{5.34a}$$
$$2u_1 - u_2 + 5u_3 = 6 \tag{5.34b}$$

$$-3u_1 + 2u_2 + u_3 = 5 \tag{5.34c}$$

may be transformed to:

$$u_2 = (5 - u_1 + 2u_3)/3 \tag{5.35a}$$
$$u_3 = (6 - 2u_1 + u_2)/5 \tag{5.35b}$$
$$u_1 = (5 - 2u_2 - u_3)/(-3) \tag{5.35c}$$

Equation (5.35) is ready to be rewritten in an iterative form:

$$u_1^{\,i} = (5 - 2u_2^{\,i-1} - u_3^{\,i-1})/(-3) \tag{5.36a}$$
$$u_2^{\,i} = (5 - u_1^{\,i-1} + 2u_3^{\,i-1})/3 \tag{5.36b}$$
$$u_3^{\,i} = (6 - 2u_1^{\,i-1} + u_2^{\,i-1})/5 \tag{5.36c}$$

where superscripts i-1 and i on the right of a variable represent that the variable is in its (i-1)-th and i-th iteration, respectively.

(2) Iteration

To start an iteration process, we have to give a guess of initial solution of all variables, i.e., the solution of variables when $i = 0$. If there is no idea about the initial solution, zeros can be assumed.

One form of the stopping or convergence criterion of an iteration might be when the largest difference of a variable is less than a given tolerance, that is,

$$\max\left(\left|u_j^{\,i} - u_j^{\,i-1}\right|\right) < tol \qquad (j = 1,2,\cdots,n) \tag{5.36d}$$

where: n = number of variables to be solved for
tol = a given tolerance

Another possible form of convergence criterion is when the sum of the squares of the differences is less than a prescribed tolerance, i.e.,

$$\sum_{j=1}^{n}(u_i^{\,i} - u_j^{\,i-1})^2 < tol \tag{5.36e}$$

Algorithm 5.7: Jacobi iteration

Given n variables: $u_1, u_2, \cdots, u_n$, an initial guess of the solution: $u_1^0, u_2^0, \cdots, u_n^0$, a maximum number of iteration steps, m, a given tolerance *tol*, and a set of iterative equations:

$$u_1^i = f_1(u_2^{i-1}, u_3^{i-1}, \cdots, u_n^{i-1})$$
$$u_2^i = f_2(u_1^{i-1}, u_3^{i-1}, \cdots, u_n^{i-1})$$
$$\vdots$$
$$u_n^i = f_n(u_1^{i-1}, u_2^{i-1}, \cdots, u_{n-1}^{i-1})$$

C..........Initialization, $NEWu_j$ ($j = 1, n$) should be real numbers.

1:        **DO** 2 $j = 1$, n
2:        $OLDu_j \leftarrow u_j^0$

C..........Calculate a new set of solution

3:        **DO** 12 $i = 1$, m
4:        **DO** 5 $j = 1$, n
5:        $NEWu_j \leftarrow f_j(OLDu_1, \cdots, OLDu_{j-1}, OLDu_{j+1}, \cdots, OLDu_n)$

C..........Convergence test

6:         $error \leftarrow 0$
7:        **DO** 8 $j = 1$, n
8:        $temp = \left| u_j^i - u_j^{i-1} \right|$
9:        **IF** ($temp > error$) $error \leftarrow temp$
10:       **IF** (error < tol) **THEN**
                    WRITE (*, *) "Answer is", $NEWu_1, NEWu_2, \cdots, NEWu_n$
                    EXIT
                ENDIF

C..........Update the solution

11:       DO 12 $j = 1$, n
12:       $OLDu_j \leftarrow NEWu_j$

C..........No convergence

13:       **WRITE** (*, *) "No convergence and Answer so far is" ,
14:       **WRITE** (*, *) $NEWu_1, NEWu_2, \cdots, NEWu_n$
15:       **END**

Table 5.1 shows the trace of the Jacobi iteration on the set of equations given in equation (3.36). Twenty-six iterations were required for the accuracy of four decimal places.

| Table 5.1   Trace of the Jacobi iteration | | | |
Iteration	u_1	u_2	u_3
0	0	0	0
1	-1.6667	1.6667	1.2000
2	-.1556	3.0222	2.2000
3	1.0815	3.1852	1.8667
4	1.0790	2.5506	1.4044
5	.5019	2.2433	1.2785
6	.2550	2.3517	1.4479
7	.3838	2.5469	1.5683
8	.5541	2.5843	1.5559
9	.5748	2.5192	1.4952
10	.5112	2.4719	1.4739
11	.4726	2.4789	1.4899
12	.4825	2.5024	1.5067
13	.5039	2.5103	1.5075
14	.5094	2.5037	1.5005
15	.5026	2.4972	1.4970
16	.4971	2.4971	1.4984
17	.4975	2.4999	1.5006
18	.5001	2.5012	1.5010
19	.5011	2.5006	1.5002
20	.5005	2.4998	1.4997
21	.4997	2.4996	1.4998
22	.4997	2.4999	1.5000
23	.5000	2.5001	1.5001
24	.5001	2.5001	1.5000
25	.5001	2.5000	1.5000
26	.5000	2.5000	1.5000

5.1.2.2 Gauss-Seidel Method

In a Gauss-Seidel iteration, equation (3.35) is transformed into another iterative form:

$$u_1^{i} = (5 - 2u_2^{i-1} - u_3^{i-1})/(-3) \tag{5.37a}$$

$$u_2^{i} = (5 - u_1^{i} + 2u_3^{i-1})/3 \tag{5.37b}$$

$$u_3^{i} = (6 - 2u_1^{i} + u_2^{i})/5 \tag{5.37c}$$

where the most recently calculated individual variables rather than the most recently calculated set of variables are used. If the coefficient of the variable solved for in each equation is larger than the sum of the absolute values of the rest of the variables, the iteration always converges using this method with any initial guess of solution. The algorithm for this method is given below.

Algorithm 5.8: Gauss-Seidel iteration

Given n variables: $u_1, u_2, \cdots, u_n$, an initial guess of the solution: $u_1^0, u_2^0, \cdots, u_n^0$, a maximum number of iteration steps, m, a given tolerance *tol*, and a set of iterative equations:

$$u_1' = f_1(u_2^{i-1}, u_3^{i-1}, \cdots, u_n^{i-1})$$
$$\vdots$$
$$u_j' = f_n(u_1^i, \cdots, u_{j-1}', u_{j+1}^{i-1}, \cdots, u_{n-1}^{i-1})$$
$$\vdots$$
$$u_n^i = f_n(u_1', u_2', \cdots, u_{n-1}')$$

```
C..........Initialization
1:          DO 2 j = 1, n
2:              OLDu_j ← u_j^0
C..........Calculate a new set of solution
3:          DO 12 i = 1, m
4:          DO 5 j = 1, n
5:              OLDu_j ← f_j(OLDu_1, ···, OLDu_{j-1}, OLDu_{j+1}, ···, OLDu_n)
C..........Convergence test
6:              error ← 0
7:          DO 8 j = 1, n
8:              temp = |u_j' - u_j^{i-1}|
9:              IF ( temp > error)    error ← temp
10:             IF (error < tol ) THEN
                        WRITE (*, *) "Answer is", OLDu_1, OLDu_2, ···, OLDu_n
                        EXIT
                ENDIF
11:         CONTINUE
C..........No convergence
12:         WRITE (*, *) "No convergence and Answer so far is" ,
13:         WRITE (*, *) OLDu_1, OLDu_2, ···, OLDu_n
14:         END
```

Table 5.2 shows the trace of the Gauss-Seidel iteration on the set of equations given in equation (3.36). Compared to Table 5.1, only fourteen iterations were required for the accuracy of four decimal places.

Table 5.2 Trace of the Gauss-Seidel iteration

Iteration	u_1	u_2	u_3
0	0	0	0
1	-1.6667	2.2222	2.3111
2	.5852	3.0123	1.5684
3	.8644	2.4241	1.3391
4	.3958	2.4275	1.5272
5	.4607	2.5312	1.5220
6	.5281	2.5053	1.4898
7	.5001	2.4932	1.4986
8	.4950	2.5007	1.5022
9	.5012	2.5010	1.4997
10	.5006	2.4996	1.4997
11	.4996	2.4999	1.5001
12	.5000	2.5001	1.5000
13	.5001	2.5000	1.5000
14	.5000	2.5000	1.5000

In this subsection, we have presented two iterative methods: Jacobi and Gauss-Seidel Iterations. One major disadvantage associated with iterative methods is that the solution method does not proceed to the answer in a fixed number of arithmetic operations. In other words, the number of arithmetic operations cannot be predicted in advance.

5.2 MATERIAL NON-LINEARITY

Material non-linearity means that the stress-strain relation of a given material is non-linear. Such non-linearity causes the difficulties, to a certain extent, in obtaining an exact solution to the differential equations which are used to describe a problem. An exact solution is obtainable with only very few simple problems, while for most problems in engineering no method solving for an exact answer exists. As a numeric method, the finite element method is an effective tool for tackling problems with material non-linearity.

In this section, only the cases of small deformation is considered. The next section will address the problems with large deformation. When the deformation of the investigated material is small,   linear equilibrium equation and geometric relation still hold as follows:

$$\int_{VOL} \mathbf{B}^T \sigma dV = \mathbf{R} \tag{5.38}$$

$$\varepsilon = \mathbf{B}\mathbf{U} \tag{5.39}$$

However, the physical equation becomes non-linear and is of the general form:

$$f(\sigma, \varepsilon) = 0 \tag{5.40}$$

Because of the non-linearity in the above equation, the relation between stress σ and displacement $\mathbf{U}$ is also non-linear, and thus equation (5.38) might be rewritten as:

$$\mathbf{K}(\mathbf{U})\,\mathbf{U} = \mathbf{R} \tag{5.41}$$

This is basically a set of non-linear equations with variables $u_1, u_2, \cdots, u_n$. Before introducing a variety of methods for solving equation (5.41), methods for solving one non-linear equation are reviewed first.

5.2.1 Methods for Solving a Non-linear Equation

In this subsection, we concentrate on the solution to equations with a single variable. For a linear equation, the solution is easily determined. For instance, the solution to the equation:

$$f(x) = 2x + 4 = 0$$

can be easily solved. For a second, third or fourth-degree polynomial equation, a solution can be expressed in a formula. However, for a fifth or higher-degree polynomial equation in a general form, no algebraic solution can be found. Furthermore, a general algebraic solution to many non-polynomial equations, such as $f(x) = \cos(x) + 3x - 1 = 0$, is not available.

Four numerical methods will be introduced below for solving non-linear equations including higher-degree polynomial ones and transcendental ones.

5.2.1.1 Iterative Substitution

This method is similar to the Jacobi method in Section 5.2.2.1. First, an original equation $f(x) = 0$ is transformed into an iterative form:

$$^i x = g(^{i-1}x) \tag{5.42}$$

where superscripts i-1 and i on the left of x represent that x is in its (i-1)-th and i-th iteration $(1 \leq i)$, respectively. Then an iterative process can proceed, provided an initial guess of solution $^0 x$ is given. The iterative process is illustrated in Fig. 5.2 in which ξ stands for the exact solution. The convergence criterion can be of the same form as in equation (5.36d) or (5.36e).

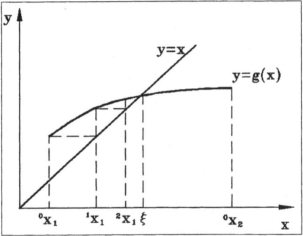

Figure 5.2   An iterative process of the iterative substitution.

One simple example of an original equation and its iterative form is:

$$f(x) = \cos(x) + 3x - 1 = 0$$
$$^i x = g(^{i-1}x) = \left[1 - \cos(^{i-1}x)\right] / 3$$

One criterion to guarantee the convergence of the iteration is when $g(x)$ is continuous and $|g'(x)| < 1$ holds for all x in the investigated domain and all intermediate variables $^1 x, ^2 x, \cdots$, where $g'(x)$ is the first derivative of g at x. In general, the convergence rate of this method is very slow.

5.2.1.2 Method of Bisection

To use the method of bisection, we need to first choose two different values of x , x_1 and x_2, such that $f(x_1) \times f(x_2) < 0$. If $f(x)$ is continuous in the domain $(x_1,$ $x_2)$, there must be a solution for f(x) = 0, ξ .

The iteration of bisection starts with calculating the midpoint $^i x_{mid}$ of $(^{i-1}x_1, ^{i-1}x_2)$:

$$^i x_{mid} = \frac{^{i-1}x_1 + ^{i-1}x_2}{2} \tag{5.43}$$

where the meaning of superscripts i and i-1 are same as in Section 3.2.1.1. If $f(^i x_{mid}) = 0$, then $^i x_{mid}$ is the solution; Otherwise, if $f(^i x_{mid})$ has the same sign as $f(^{i-1}x_k)$ (k = 1 or 2), then assign $^i x_k$ and $^i x_{3-k}$ with the values of $^i x_{mid}$ and $^{i-1}x_{3-k}$, respectively. Continue the iteration until a solution is found or a convergence criterion is satisfied or the maximum number of iterations has been reached.

With the exception of the convergence criteria in equations (5.36d) or (5.36e), $f(x_{mid})$ can also be used as an alternative test of convergence. If it falls into a vicinity of zero specified by a given tolerance, the process has converged. The iterative process of this method is illustrated in Fig. 5.3 and formalized by the following algorithm.

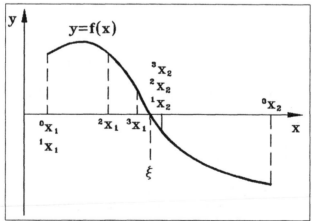

Figure 5.3 An iterative process of the successive bisection.

Algorithm 5.9: Method of bisection
Given a non-linear equation $f(x) = 0$, two values x_1 and x_2 ($x_1 < x_2$) such that
$f(x)$ is continuous in the domain (x_1, x_2) and $f(x_1) \times f(x_2) < 0$, a maximum
number of iteration steps, m and a given tolerance tol.

```
C.........Begin an iteration
1:       DO 5  i = 1, m
C.........Find the midpoint
```

2:          $x_{mid} \leftarrow \dfrac{x_1 + x_2}{2}$

```
C.........Check the termination conditions
```

3:          **IF** ($\left| f(x_{mid}) \right| < tol$ **.OR.** $\left| x_1 - x_2 \right| < tol$) **WRITE**(*,*) x_{mid} , $f(x_{mid})$

```
C.........Bisect the domain (x₁, x₂)
```
4:          **IF** ($f(x_{mid}) \times f(x_1) < 0$) **THEN**

                $x_2 \leftarrow x_{mid}$

       ELSE

                $x_1 \leftarrow x_{mid}$

```
5:       ENDIF
6:       WRITE (*,*) "No convergence, solution so far is", x_mid
7:       END
```

Table 5.4 shows the results of the bisection method solving an equation
$f(x) = x^3 + 2x^2 - 3x - 7 = 0$ with two starting points: $x_1 = 1.8, x_2 = 1.9$. The
second and third columns represent the value of x_{mid} and $f(x_{mid})$, respectively.
To achieve an accuracy of six digits to the decimal point, 17 iterations are
required.

5.2.1.3 Secant Method

The secant method is another iterative approach which, in general, converges
more quickly than the method of successive bisection. It also requires two
points, x_1 and x_2 , before starting an iterative process. To definitely control an
iterative process, the restriction on the function $f(x)$ is that in the domain (x_1, x_2),
$f(x)$ is continuous and the signs of both $f'(x)$ and $f''(x)$ do not change, where
$f'(x)$ and $f''(x)$ are the first and second derivative of f at x, respectively.

The basic idea behind the secant method is that a portion of a curve of a
equation between two points, x_1 and x_2, is approximated by a straight line (or a
secant line) connecting these two points; The intersection point, x_{new}, between
the secant line and x axis is the approximation to the solution. Replace one of

x_1 and x_2 with x_{new} and a new iteration can proceed on a new pair of x_1 and x_2. An iterative process continues until a solution is found or a convergence criterion is satisfied or the maximum number of iterations has been reached.

Table 5.4 Trace of the bisection method

No. of iteration, i	$^i x$	$f(^i x) = {}^i x^3 + 2\,^i x^2 - 3\,^i x - 7 = 0$
0	$x_1 = 1.8,\ x_2 = 1.9$	/
1	1.850000	.6266240
2	1.825000	.2646396
3	1.812500	.0871578
4	1.806250	-.0007110
5	1.809375	.0431515
6	1.807812	.0212019
7	1.807031	.0102403
8	1.806641	.0047637
9	1.806445	.0020267
10	1.806348	.0006576
11	1.806299	-.0000274
12	1.806323	.0003147
13	1.806311	.0001430
14	1.806305	.0000574
15	1.806302	.0000152
16	1.806300	-.0000061
17	1.806301	.0000052

Let $x_1 < x_2$ and $f(x_1) \times f(x_2) < 0$. When $f'(x)f''(x) > 0$ as shown in Fig. 5.4 (a), it is easy to prove that the following formula is valid:

$$\begin{cases} ^i x_{new} = {}^{i-1}x_2 - \dfrac{{}^{i-1}x_2 - {}^{i-1}x_1}{f({}^{i-1}x_2) - f({}^{i-1}x_1)} \times f({}^{i-1}x_2) \\ ^i x_1 \leftarrow {}^i x_{new} \quad ^i x_2 \leftarrow {}^{i-1}x_2 \end{cases} \tag{5.44a}$$

where the meaning of superscripts i and i-1 are same as in Section 3.2.1.1. When $f'(x)f''(x) < 0$ as illustrated in Fig. 5.4 (b), a different iterative formula applies:

$$\begin{cases} {}^{i}x_{new} = {}^{i-1}x_1 - \dfrac{{}^{i-1}x_2 - {}^{i-1}x_1}{f({}^{i-1}x_2) - f({}^{i-1}x_1)} \times f({}^{i-1}x_1) \\ {}^{i}x_2 \leftarrow {}^{i}x_{new} \qquad {}^{i}x_1 \leftarrow {}^{i-1}x_1 \end{cases}$$

(5.44b)

Figure 5.4 illustrates the process of convergence in both cases.

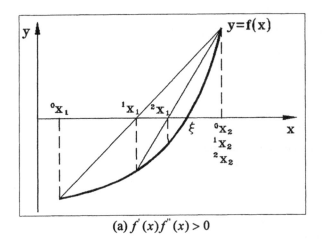

(a) $f'(x)f''(x) > 0$

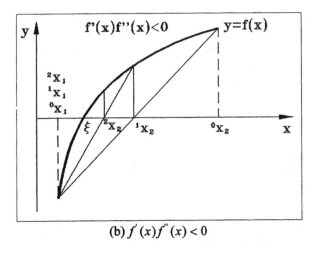

(b) $f'(x)f''(x) < 0$

Figure 5.4 Approximation of a solution using the secant method.

Algorithm 5.10: Secant Method

Given a non-linear equation $f(x) = 0$, two values x_1 and x_2 ($x_1 < x_2$ and $f(x_1) \times f(x_2) < 0$) such that $f(x)$ is continuous and the signs of both $f'(x)$ and $f''(x)$ do not change in the domain (x_1, x_2), a maximum number of iteration steps, m and a given tolerance *tol*.

C..........Calculate a flag

1:      **IF**($f'(x_1)f''(x_1) > 0$) **THEN**

            Iflag = 1

        ELSE

            Iflag = 0

        ENDIF

C..........Begin an iteration

2:      **DO 7** $i = 1, m$

C..........Calculate a new estimate

3:      $x_{new} \leftarrow \dfrac{x_2 + x_1}{f(x_2) - f(x_1)}$

4:      **IF** (*Iflag* = 1) **THEN**

            $x_{new} \leftarrow x_2 - x_{new} \times f(x_2)$

        ELSE

            $x_{new} \leftarrow x_1 - x_{new} \times f(x_1)$

        ENDIF

C..........Check the termination conditions

5:      **IF** ($\left| f(x_{new}) \right| < tol$.**OR**. $\left| x_1 - x_2 \right| < tol$) **WRITE**(*,*) $x_{new}, f(x_{new})$

C..........Update variables

6:      **IF** (*Iflag* = 1) **THEN**

            $x_1 \leftarrow x_{new}$

        ELSE

            $x_2 \leftarrow x_{new}$

7:      **ENDIF**

8:      **WRITE** (*,*) "No convergence, solution so far is", x_{new}

9:      **END**

Table 5.5 gives the result of the secant method solving the same equation as in the previous subsection with two starting points: $x_1 = 1.8, x_2 = 1.9$. The second and third columns represent the value of x_{new} and $f(x_{new})$, respectively.

Iteration steps required are reduced from 17 to 5, compared to the successive bisection method.

Table 5.5 Trace of the secant method

No. of iteration, i	^ix	$f(^ix)=^ix^3 +2^ix^2 -3^ix - 7 = 0$
0	$x_1 = 1.8, x_2 = 1.9$	/
1	1.805999	-.0042318
2	1.806286	-.0002032
3	1.806300	-.0000094
4	1.806301	-.0000011
5	1.806301	.0000006

5.2.1.4 Newton-Raphson Method

The basic idea of the Newton-Raphson method is that a portion of curve of a equation starting from one point x_1 is approximated by a tangent to the curve at point x_1. The intersection point, x_{new}, between the tangent line and x axis is the approximation to the solution. Assign x_1 with the value of x_{new} and a new iteration can proceed. An iterative process continues until a solution is found or a convergence criterion is satisfied or the maximum number of iterations has been reached.

The Newton-Raphson method requires only one starting point x_1. However, to definitively control an iterative process, the restriction on the function $f(x)$ is that in a given domain (x_1, x_2), $f(x)$ and $f'(x)$ are continuous, and both $f'(x)$ and $f''(x)$ are not equal to zero, where $f'(x)$ and $f''(x)$ are the first and second derivative of f at x, respectively.

The tangent to a curve at point x_1 in Fig. 5.5 can be expressed by:

$$\frac{y - f(x_1)}{x - x_1} = f'(x_1)$$

The x-axis intersection of this tangent is:

$$x = x_1 - \frac{f(x_1)}{f'(x_1)}$$

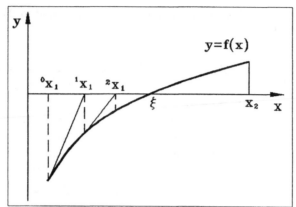

Figure 5.5 Successive approximation to the solution using Newton-Raphson method.

which can be written in a general form:

$$^{i}x = {}^{i-1}x - \frac{f(^{i-1}x)}{f'(^{i-1}x)}$$ (5.45)

where: i = number of iteration steps and $1 \leq i$.

Successive approximations are made until a convergence criterion is satisfied or the maximum number of iterations has been reached.

Algorithm 5.11: Newton-Raphson method
Given a non-linear equation $f(x) = 0$, two values x_1 and x_2 ($x_1 < x_2$) such that $f(x)$ and $f'(x)$ are continuous and both $f'(x)$ and $f''(x)$ are not equal to zero in the domain (x_1, x_2), a maximum number of iteration steps, m and a given tolerance *tol*.

C.........Begin an iteration
1:        **DO** 4 $i = 1, m$
C.........Calculate a new estimate
2:        $x_{new} \leftarrow x_1 - \dfrac{f(x_1)}{f'(x_1)}$
C.........Check the termination conditions
3:        **IF** ($|f(x_{new})| < tol$.**OR.** $|x_{new} - x_1| < tol$) **WRITE**(*,*) x_{new} , $f(x_{new})$

C..........Update variables
4:          $x_1 \leftarrow x_{new}$
5:          **WRITE** (*,*) "No convergence, solution so far is", x_{new}
6:          **END**

Table 5.6 shows the results using the Newton-Raphson method solving the same equation as in Section 5.2.1.3 with one starting point: $x_1 = 1.8$. The number of iteration steps required is even less than that required for the secant method to achieve an accuracy of six digits to the decimal point.

Table 5.6   Trace of Newton-Raphson method

No. of iteration, i	^ix	$f(^ix)=^ix^3 +2^ix^2 -3^ix - 7 = 0$
0	1.800000	-.0880004
1	1.806322	-.0880008
2	1.806301	.0002961
3	1.806301	.0000007

5.2.1.5 Summary

Among the four methods introduced in the previous subsections,   the Newton-Raphon method converges fastest, followed by the secant method, bisection method and iterative substitution. As to the numerical stability, the method of bisection behavior seems to be the best. It requires only the continuity of the function f(x) in the investigated domain, while the secant and Newton-Raphson method require some restriction on the first and second derivatives of f at x. For the iterative substitution, $|g'(x)| < 1$ should hold to guarantee that the equation is solvable.

5.2.2 Methods for Solving Non-linear Elastic Problems

Generally, there are three basic types of methods available for solving non-linear elastic problems.

5.2.2.1 Method of Variable Stiffness
1. Secant stiffness (or Direct iteration)
If the stress-strain relation of a soil can be expressed in the following form:

$$\sigma = C^e(\varepsilon)\varepsilon \tag{5.46}$$

then based upon equation (5.39), the above equation can be rewritten as:

$$\sigma = \mathbf{C}^e(\varepsilon)\mathbf{BU} = \mathbf{C}^e(\mathbf{U})\mathbf{BU} \tag{5.47}$$

Substituting equation (5.46) into equation (5.38) yields:

$$\mathbf{K}(\mathbf{U}) = \int_{VOL} \mathbf{B}^T \mathbf{C}^e(\mathbf{U})\mathbf{B}dV \tag{5.48}$$

We can transform Eq.(5.41) into an iterative form:

$${}^{i-1}\mathbf{K}\,{}^i\mathbf{U} = \mathbf{R} \tag{5.49}$$

The iterative process of equation (5.49) is explained below. First, let ${}^0\mathbf{U} = 0$ and $\mathbf{K}({}^0\mathbf{U}) = {}^0\mathbf{K}$. Substituting ${}^0\mathbf{K}$ into equation (5.49) leads to:

$${}^1\mathbf{U} = {}^0\mathbf{K}^{-1}\mathbf{R}$$

where: ${}^1\mathbf{U}$ = the first approximation to the solution of displacement variables

Based on the ${}^1\mathbf{U}$, using equations (5.39), (5.46) and (5.48) can determine ${}^1\mathbf{K}$. By the substitution of ${}^1\mathbf{K}$ into equation (5.49), ${}^2\mathbf{U}$ is solved as the second approximation to displacement variables. Repeat such iterations until ${}^n\mathbf{U} \cong {}^{n-1}\mathbf{U}$, where ${}^n\mathbf{U}$ is considered as the solution to the non-linear equations in equation (5.41).

Fig. 5.6 illustrates such an iterative process in which the stress of an element gradually approaches the true value. From this figure, it can be seen that the elastic matrix $\mathbf{C}^e(\varepsilon)$ denotes the slope of secant lines on a stress-strain curve. This is the reason why the method is called secant stiffness method or direct iteration method.

2. Tangential stiffness
If the stress-strain relation of a soil can be expressed in an incremental form:

$$d\sigma = \mathbf{C}_T^e(\varepsilon)d\varepsilon \tag{5.50}$$

then the tangential stiffness method can be applied, where $\mathbf{C}_T^e(\varepsilon)$ is a tangential elastic matrix.

Equation (5.38) may be rewritten as:

$$\Psi(\mathbf{U}) = \int_{VOL} \mathbf{B}^T \sigma dV - \mathbf{R} = 0 \tag{5.51}$$

Now calculate the change of Ψ caused by an increment $d\mathbf{U}$. Because $\mathbf{R}$ is independent on $\mathbf{U}$, therefore,

$$d\Psi = \int_{VOL} \mathbf{B}^T d\sigma \, dV \tag{5.52}$$

Substituting equation (5.50) into the above equation and considering equation (5.39) leads to:

$$d\Psi = \left(\int_{VOL} \mathbf{B}^T \mathbf{C}_T^e(\varepsilon) \mathbf{B} \, dV \right) d\mathbf{U} = \mathbf{K}_T^e d\mathbf{U} \tag{5.53}$$

where: $\mathbf{K}_T^e$ = a tangent stiffness matrix which is of the form:

$$\mathbf{K}_T^e = \int_{VOL} \mathbf{B}^T \mathbf{C}_T^e(\varepsilon) \mathbf{B} \, dV \tag{5.54}$$

Using the Newton-Raphson method introduced in Section 5.2.1.4, an iterative formula can be obtained as follows:

$$
\begin{aligned}
{}^i\mathbf{K}_T^e \, \Delta^{i+1}\mathbf{U} &= -{}^i\Psi \\
{}^{i+1}\mathbf{U} &= {}^i\mathbf{U} + \Delta^{i+1}\mathbf{U}
\end{aligned}
\tag{5.55}
$$

where:

$$
{}^i\Psi = \int_{VOL} \mathbf{B}^T \, {}^i\sigma \, dV - \mathbf{R} \tag{5.56}
$$

The iterative process of this method is explained as follows. If the i-th approximation of displacement ${}^i\mathbf{U}$ is known, ${}^i\varepsilon$ is determined by equation (5.39). Then the tangent elastic matrix $\mathbf{C}_T^e({}^i\varepsilon)$ can be determined using the incremental relation between stress and strain. Substituting $\mathbf{C}_T^e({}^i\varepsilon)$ into equation (5.54), the tangent matrix ${}^i\mathbf{K}_T^e = \mathbf{K}_T^e({}^i\mathbf{U})$ is solved. From equation

(5.56), $^i\Psi$ is calculated. By the substitution of $'\mathbf{K}_T^e$ and $^i\Psi$ into equation
(5.55), a set of linear equations can be solved to obtain $\Delta^{i+1}\mathbf{U}$. Consequently,
the (i+1)-th approximation to displacement is $^{i+1}\mathbf{U}=^i\mathbf{U}+\Delta^{i+1}\mathbf{U}$. Repeat the
above process until a convergence is achieved.

The changes of stress in an element during an iterative process of this method
is illustrated in Fig. 5.6.

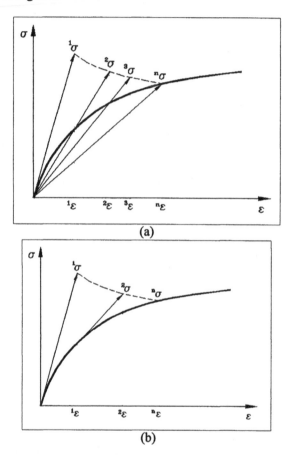

Figure 5.6 Change of stress in an element during an iterative process.

5.2.2.2 Method of Initial Stress

If the physical equation of a soil is of the form:

$$\sigma = f(\varepsilon) \tag{5.57}$$

the value of a stress can be determined by a given strain. We presume that the above equation can be replaced by a linear elastic equation with an initial stress:

$$\sigma = \mathbf{C}^e \, \varepsilon + \sigma_0 \tag{5.58}$$

where: σ_0 = a column matrix of initial stress

$\mathbf{C}^e$ = a linear elastic matrix which is equal to the tangent elastic matrix of a non-linear soil, $\mathbf{C}_T^e(\mathbf{U})$, at $\mathbf{U} = 0$

Adjust the initial stress σ_0 such that for a given strain ε, the σ predicted by equation (5.57) is equal to that by equation (5.58). This requirement gives:

$$\sigma_0 = \sigma - \mathbf{C}^e \, \varepsilon = f(\varepsilon) - \mathbf{C}^e \, \varepsilon \tag{5.59}$$

Assume a fictitious linear elastic stress $\sigma^{el} = \mathbf{C}^e \varepsilon$. Then, as illustrated in Fig. 5.7, the above equation is rewritten as:

$$\sigma_0 = -(\sigma^e - \sigma) \tag{5.60}$$

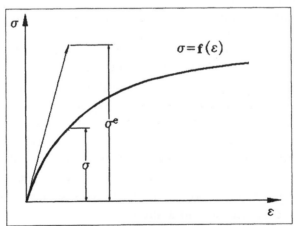

Figure 5.7 The definition of an initial stress.

Substituting equation (5.58) into equation (5.38) yields:

$$\left(\int_{VOL} \mathbf{B}^T \mathbf{C}^e \mathbf{B} \, dV \right) \mathbf{U} = \mathbf{R} - \int_{VOL} \mathbf{B}^T \sigma_0 \, dV \qquad (5.61)$$

Let $\mathbf{K}^e = \int_{VOL} \mathbf{B}^T \mathbf{C}^e \mathbf{B} \, dV$, equation (5.61) can be rewritten as:

$$\mathbf{K}^e \mathbf{U} = \mathbf{R} - \int_{VOL} \mathbf{B}^T \sigma_0 \, dV \qquad (5.62)$$

where, $\mathbf{K}^e$ is the global stiffness matrix defined by a linear elastic matrix. The above equation can be transformed into an iterative form:

$$\mathbf{K}^{e \; i+1}\mathbf{U} = \mathbf{R} - {}^i\mathbf{R}$$
$$ {}^i\mathbf{R} = \int_{VOL} \mathbf{B}^T ({}^i\sigma - {}^{i-1}\sigma^e) dV \qquad (5.63)$$

If the i-th approximation to displacement ${}^i\mathbf{U}$ is known, ${}^i\sigma$ is determined by equations (5.39) and (5.57). Using known $\mathbf{C}^e$ and ${}^i\varepsilon$, ${}^i\sigma^e$ is calculated. Based on the second equation in equation (5.63), ${}^i\mathbf{R}$ is solved; the first equation can be used to determine ${}^{i+1}\mathbf{U}$. Repeat the above process until a convergence is encountered. The value of ${}^0\mathbf{R}$ is usually assumed to be zero, which corresponds to the solution of an elastic problem.

Since the stiffness matrix $\mathbf{K}^e$ does not change in an entire iterative process, this method is also called the constant stiffness method.

The change in stress and strain in an element is shown in Fig. 5.8 in which σ_m and ε_m are the true values of stress and strain. From this figure, it can be seen that if $\sigma_0 = \sigma_m - \sigma_m^e$ is used as the initial stress of an element, then the predictions by the linear relation in equation (5.58) and the non-linear relation in equation (5.57) are consistent. Therefore, an entire iterative process is equivalent to the process adjusting the initial stress in an element. ${}^i\mathbf{R}$ is the equivalent nodal forces corresponding to an initial stress. Once the initial stress has been adjusted to $\sigma_0 = \sigma_m - \sigma_m^e$, the linear elastic solution with an initial stress σ_0 is the same as the original non-linear elastic solution. Therefore, this method is called the method of initial stress.

5.2.2.3 Method of Initial Strain

In some cases such as creep, the element strain is expressed by element stress as follows:

$$\varepsilon = f(\sigma) \tag{5.64}$$

Similar to the assumption made in the initial stress method, we presume that the above equation can be substituted by a linear elastic physical equation with an initial strain:

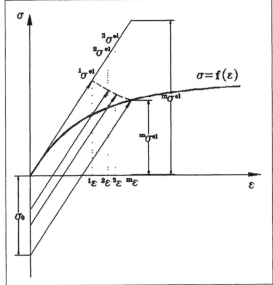

Figure 5.8 Change in element stress in an iterative process of the initial stress method.

$$\sigma = \mathbf{C}^e (\varepsilon - \varepsilon_0) \tag{5.65}$$

where, ε_0 is an initial strain. For a given stress σ, adjust ε_0 such that the values of strains predicted by the above two equations are the same, that is,

$$\varepsilon_0 = \varepsilon - (\mathbf{C}^e)^{-1}\sigma = f(\sigma) - (\mathbf{C}^e)^{-1}\sigma \tag{5.66}$$

After introducing a fictitious linear elastic strain $\varepsilon^e = (C^e)^{-1}\sigma$, the above equation can be rewritten as:

$$\varepsilon_0 = \varepsilon - \varepsilon^e \tag{5.67}$$

Fig. 5.9 shows the relationship defined by equation (5.67). Substituting equation (5.65) into equation (5.38) yields:

$$K^e U = R + \int_{VOL} B^T C^e \varepsilon_0 \, dV \tag{5.68}$$

which can be written in an iterative form:

$$K^e \,^{i+1}U = R + {}^i R$$
$${}^i R = \int_{VOL} B^T C^e \,{}^i \varepsilon_0 dV \tag{5.69}$$

where: ${}^i R$ = a correction load which approximates the load required for the initial strain

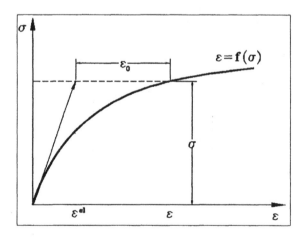

Figure 5.9 The definition of an initial strain.

The iterative process of this method is as follows. If the *i*-th approximation to the displacement variables is known, ${}^i \varepsilon$ is calculated using equation (5.39).

Based on $'\varepsilon$ and the initial strain obtained from the last iteration, $^{i-1}\varepsilon_0$, the value of element stress can be determined using equation (5.65) as follows:

$$'\sigma = \mathbf{C}^e('\varepsilon - {}^{i-1}\varepsilon_0)$$ (5.70)

Then, using equation (5.64), the value of strain $'\varepsilon^\sigma$ corresponding to $'\sigma$ is solved. Thus, the initial strain is determined by

$$^i\varepsilon_0 = {}^i\varepsilon^\sigma - {}'\varepsilon^e = {}'\varepsilon^\sigma - (\mathbf{C}^e)^{-1}\,{}^i\sigma$$ (5.71)

Substituting the above equation into equation (5.69) and solving it will lead to the solution to the $(i+1)$-th approximation to the displacement $^{i+1}\mathbf{U}$. Repeat the above process until a convergence occurs. Fig. 5.10 shows the change of stress and strain in an element during an iterative process of the initial strain method.

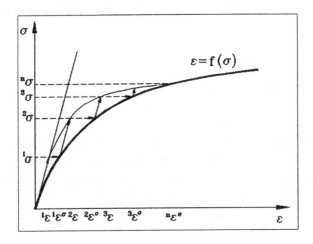

Figure 5.10   Change of element stress in an iterative process of the initial strain method.

5.2.3 Methods for Solving Non-linear Elasto-Plastic Problems

It is assumed that the elasto-plastic strain increment and stress increment can be expressed approximately by the following relation:

$$\Delta\sigma = \mathbf{C}^{ep}\Delta\varepsilon$$ (5.72)

where, $\mathbf{C}^{ep}$ is an elasto-plastic matrix which is a function of the current stress level but independent upon the increments of stress and strain. Thus, the above equation can be viewed as a linear relation. Some forms of $\mathbf{C}^{ep}$ have been derived in Section 3.3.4.1.

For the purpose of linearization, a load increment method is used, that is, load increments are added step by step; in each step, a small enough load increment is applied at the current stress and strain levels such that a non-linear problem can be approximated by a series of linear problems.

In a loading process, an elasto-plastic problem is essentially the same as an equivalent non-linear elastic problem. Thus, the methods introduced in the last section can be applied to elasto-plastic problems. In an unloading process, $\mathbf{C}^{ep}$ should be replaced by an elastic matrix $\mathbf{C}^{e}$, which implies that the unloading process is a linear elastic problem. Backing to the loading process, in the methods discussed in the last section, a total load is applied just one time followed by an iterative process, while in dealing with an elasto-plastic loading process a total load is divided into many increments and each time only one increment is added. Below three types of load increment methods will be introduced.

5.2.3.1 Incremental Tangent Stiffness
At the beginning of a loading process, an object is still in its elastic state and thus an linear elastic matrix $\mathbf{C}^{e}$ can be used in the calculation of stress and strain. After some elements of this object enter their plastic states, an incremental load method should be used.

Let $^{0}\mathbf{U}$, $^{0}\varepsilon$ and $^{0}\sigma$ be, respectively, the displacement, strain and stress at the moment when at least one element of the object begins to enter its plastic state. On the basis of these initial values, let an load increment $\Delta^{1}\mathbf{R}$ be added. For those elements which are still in elastic state, their element stiffness matrices are of the form:

$$\mathbf{k} = \int_{VOL} \mathbf{B}^{T} \mathbf{C}^{e} \mathbf{B} dV \tag{5.73a}$$

while for the elements in plastic state, the element stiffness matrix is of the form:

$$\mathbf{k} = \int_{VOL} \mathbf{B}^{T} \mathbf{C}^{ep} \mathbf{B} dV \tag{5.73b}$$

where, the stress in $\mathbf{C}^{ep}$ should be assigned by the current stress level $^0\sigma$. All element stiffness matrices can be assembled to form a global stiffness matrix $^0\mathbf{K}$ which is dependent upon the current stress level.

Solving the following equilibrium equation:

$$^0\mathbf{K}\Delta^1\mathbf{U}=\Delta^1\mathbf{R}$$

we obtain $\Delta^1\mathbf{U}$, and then $\Delta^1\varepsilon$ and $\Delta^1\sigma$. Therefore, the new levels of displacement, strain and stress after the first load increment are:

$$^1\mathbf{U}=^0\mathbf{U}+\Delta^1\mathbf{U}$$
$$^1\varepsilon=^0\varepsilon+\Delta^1\varepsilon$$
$$^1\sigma=^0\sigma+\Delta^1\sigma$$

Repeat the above calculations until the applying of a series of load increments is completed. This iterative process may be written in a general form:

$$^{i-1}\mathbf{K}\Delta^i\mathbf{U}=\Delta^i\mathbf{R}$$

and

$$^i\mathbf{U}=^{i-1}\mathbf{U}+\Delta^i\mathbf{U}$$
$$^i\varepsilon=^{i-1}\varepsilon+\Delta^i\varepsilon$$
$$^i\sigma=^{i-1}\sigma+\Delta^i\sigma$$

The displacement, strain and stress obtained in the final iteration are the solution to the elasto-plastic problem.

In general, the plastic zone inside an object gradually expands in an incremental loading process. Although some elements are still in elastic state, they may be adjacent to the plastic zone and will enter the plastic zone after the current incremental loading step. The area constituted by these elements are called the transient zone.

For the elements in the transient zone, since their states will change from elastic to plastic after the current incremental loading step, a large error of calculating an element stiffness matrix would be incurred using either equation (5.73a) or equation (5.73b). In addition, for the elements which experience an unloading and then a reloading, a similar situation would occur. Fig. 5.11 illustrates a large error in predicting $\Delta\sigma$ when stress is considered changing from point A to point B at the end of a reloading process. To reduce the error of predicting $\Delta\sigma$, the stress increment is calculated using the following equation:

$$\Delta\sigma = \int_0^{m\Delta\varepsilon} \mathbf{C}^e \, d\varepsilon + \int_{m\Delta\varepsilon}^{\Delta\varepsilon} \mathbf{C}^{ep} \, d\varepsilon = \int_0^{\Delta\varepsilon} \mathbf{C}^e \, d\varepsilon - \int_{m\Delta\varepsilon}^{\Delta\varepsilon} \mathbf{C}^p \, d\varepsilon \qquad (5.74a)$$

where, $m\Delta\varepsilon$ is a portion of strain increment corresponding to the interval prior to the reoccurrence of plastic deformation. In order to determine m, first calculate the portion of strain increment $\Delta\bar{\varepsilon}_c$ required for stress to reach its yield point, and then estimate the equivalent strain increment $\Delta\bar{\varepsilon}_{es}$ caused by the load increment. The expression of m is of the form:

$$m = \frac{\Delta\bar{\varepsilon}_c}{\Delta\bar{\varepsilon}_{es}} \qquad (0 < m < 1) \qquad (5.74b)$$

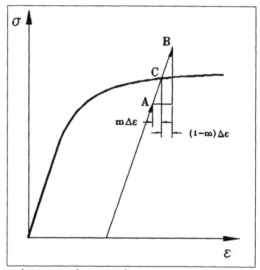

Figure 5.11 Stress increment in a transient zone.

If the load increment is small enough, equation (5.74a) can be approximated by:

$$\Delta\sigma = \left(m\mathbf{C}^e - (1-m)\mathbf{C}^{ep} \right)\Delta\varepsilon = (\mathbf{C}^e - (1-m)\mathbf{C}^p)\Delta\varepsilon \qquad (5.74c)$$

from which we define an average elasto-plastic matrix as follows:

$$\overline{\mathbf{C}}^{ep} = m\mathbf{C}^{e} - (1-m)\mathbf{C}^{ep} \tag{5.75}$$

Thus, for elements in the transient zone or at the end of a reloading process, their stiffness matrices should be calculated by:

$$\mathbf{k} = \int_{VOL} \mathbf{B}^{T}\overline{\mathbf{C}}^{ep}\mathbf{B}dV \tag{5.76}$$

Usually, the first estimation of $\Delta\overline{\varepsilon}_{es}$ is not accurate enough, because we consider the element in elastic state. The calculated results can be fed back to modify $\Delta\overline{\varepsilon}_{es}$. After several iterations, quite an accurate solution can be obtained.

It should be noted that $\Delta\varepsilon$ calculated by equation (5.72) is only the approximation to the true solution, because the incremental relation between stress and strain ought to be expressed by infinitesimal increments rather than finitesimal increments. Therefore, only when load increments are small enough will the solution be close to the true answer. A more accurate calculation may result from the combination of the incremental method introduced here and an iterative method such as Newton-Raphson method.

5.2.3.2 Incremental Initial Stress
For an elasto-plastic problem, the stress-strain relation in an incremental form may be defined as:

$$d\sigma = \mathbf{C}^{e}d\varepsilon + d\sigma_{0} \tag{5.77a}$$

and

$$d\sigma_{0} = -\mathbf{C}^{p}d\varepsilon \tag{5.77b}$$

where:

$$\mathbf{C}^{p} = \mathbf{C}^{e} - \mathbf{C}^{ep} \tag{5.77c}$$

$d\sigma_{0}$ is equivalent to the initial stress in a linear elastic problem. The infinitesimal variables in the above equation can be approximated by:

$$\Delta\sigma = \mathbf{C}^{e}\Delta\varepsilon + \Delta\sigma_{0}$$
$$\Delta\sigma_{0} = -\mathbf{C}^{p}\Delta\varepsilon \tag{5.78}$$

The equilibrium equation which should be satisfied by the displacement increment ΔU is:

$$\mathbf{K}^e \Delta \mathbf{U} = \Delta \mathbf{R} + \mathbf{R}(\Delta \varepsilon) \qquad (5.79a)$$

where:

$$\mathbf{K}^e = \int_{VOL} \mathbf{B}^T \mathbf{C}^e \mathbf{B} dV \qquad (5.79b)$$

and

Algorithm 5.12: Incremental Tangent Stiffness
Given a total load $\mathbf{R}$ applying to an investigated object, and the number of loading increments n.

1:      Apply the total load $\mathbf{R}$ to the object and conduct a linear elastic calculation:

$$\mathbf{K}^e \mathbf{U} = \mathbf{R}$$

where, $\mathbf{K}^e$ is a global stiffness matrix assembled from all element stiffness matrices of the form in equation (5.73a)

2:      Solve the equivalent stresses in each element and assign the maximum value to $\overline{\sigma}_{max}$. If $\overline{\sigma}_{max} < \sigma_f$, where σ_f is a yielding stress, the result of the elastic analysis is the solution; otherwise, let $L = \overline{\sigma}_{max} / \sigma_f$ and store the results at the load $\dfrac{1}{L}\mathbf{R}$. $\Delta \mathbf{R} = \dfrac{1}{n}\left(1 - \dfrac{1}{L}\right)\mathbf{R}$ is used as a load increment for each n iteration.

C..........Begin an incremental loading process
3:      **DO** 10 $i = 1, n$
4:      Apply a load increment $\Delta \mathbf{R}$, estimate the strain increment $\Delta \overline{\varepsilon}_{es}$ in each element and determine m using equation (5.74b).
C..........Begin an iterative process for correcting the calculation of stresses in elements in a transient zone or at the end of a reloading process
5:      **DO** 8 $j = 1, 3$
6:      Form the element stiffness matrix based upon the cases where the element is in an elastic, plastic or transient zone.

$$
{}^{i-1}\mathbf{k} = \begin{cases} \displaystyle\int_{VOL} \mathbf{B}^T \mathbf{C}^e \mathbf{B} dV & \text{in elastic state} \\[2mm] \displaystyle\int_{VOL} \mathbf{B}^T \mathbf{C}^{ep} \mathbf{B} dV & \text{in plastic state} \\[2mm] \displaystyle\int_{VOL} \mathbf{B}^T \overline{\mathbf{C}}^{ep} \mathbf{B} dV & \text{in transient zone} \end{cases}
$$

7:        Assemble all element stiffness matrices into a global stiffness matrix ${}^{i-1}\mathbf{K}$.

8:        Solve the following equilibrium equation to obtain displacement increment and then strain increment. According to these increments, modify $\Delta\overline{\varepsilon}_{es}$ and m.

$$
{}^{i-1}\mathbf{K}\Delta^i\mathbf{U} = \Delta^i\mathbf{R}
$$

C..........Update information after each increment step

9:        Calculate stress increments and superimpose the displacement, strain and stress increments to the previous levels.

$$
{}^i\mathbf{U} = {}^{i-1}\mathbf{U} + \Delta^i\mathbf{U}
$$

$$
{}^i\varepsilon = {}^{i-1}\varepsilon + \Delta^i\varepsilon
$$

$$
{}^i\sigma = {}^{i-1}\sigma + \Delta^i\sigma
$$

10:      Output the calculated results

11:      **END**

$$
\overline{\mathbf{R}}(\Delta\varepsilon) = \int_{VOL} \mathbf{B}^T \mathbf{C}^p \Delta\varepsilon \, dV \tag{5.79c}
$$

The $\overline{\mathbf{R}}(\Delta\varepsilon)$ in the above equation denotes the equivalent nodal forces (or correction load) transformed from the initial stress $\Delta\sigma_0$.

It is noted that the second term on the right hand side of equation (5.79a) is dependent upon the strain increment $\Delta\varepsilon$, but $\Delta\varepsilon$ is an undetermined variable which will become known after solving equation (5.79a). Therefore, for each load increment, an iterative process is needed to obtain displacement and strain increments. The method of increment initial stress is actually a combination of incremental and iterative methods.

The iterative equation for the i-th load increment is:

$$\mathbf{K}^e \Delta^i \mathbf{U}^{(j)} = \Delta^i \mathbf{R} + {}^i \overline{\mathbf{R}}^{(j-1)} \qquad (j = 0, 1, \cdots) \qquad (5.80)$$

where, the superscripts i and j correspond to the number of incremental and iterative steps, respectively. If the $(j-1)$-th approximation of strain increment $\Delta^i \varepsilon^{(j-1)}$ is known, equation (5.78) can be used to obtain the $(j-1)$-th approximation of stress increment $\Delta' \sigma^{(j-1)}$ based on the current stress level. Then using (5.79b) yields the correction load ${}^i \overline{\mathbf{R}}^{j-1}$ and solving equation (5.80) produces $\Delta^i \varepsilon^{(j)}$ which can be used to begin a new iteration. Repeat the iterative process until the difference of strain increments in two consecutive iterations is less than a prescribed tolerance. At this moment, we superimpose the strain and stress increments to the current strain and stress level and conduct a succeeding loading step until all loads are applied.

It should be noted that for the elements in a transient zone, the calculation of initial stress should not include the portion of $\Delta \varepsilon$ which corresponds to the interval prior to the yielding point. If a load increment is sufficiently small, using equation (5.74c), the correction load can be calculated by:

$$\overline{\mathbf{R}} = \int_{VOL} \mathbf{B}^T \mathbf{C}^P (1-m) \Delta \varepsilon \, dV \qquad (5.81)$$

where, $\mathbf{C}^P$ is determined by equation (5.77c).

5.2.3.3 Incremental Initial Strain

For an elasto-plastic problem, the stress-strain relation in an incremental form may be defined by:

$$d\sigma = \mathbf{C}^e d\varepsilon^e = \mathbf{C}^e (d\varepsilon - d\varepsilon_0) \qquad (5.82)$$

where:

$$d\varepsilon_0 = d\varepsilon^P$$

According to equations (3.49) and (3.57), we have:

$$d\varepsilon^P = \beta \frac{\partial g(\sigma)}{\partial \sigma} = \mathbf{L}^{-1} \left(\frac{\partial \mathbf{f}}{\partial \sigma} \right)^T \frac{\partial g(\sigma)}{\partial \sigma} d\sigma \tag{5.83}$$

$d\varepsilon_0$ is equivalent to an initial strain in a linear elastic problem. The infinitesimal variables in equations (15.82) and (15.83) can be approximated by finitesimal variables as follows:

$$\Delta\sigma = \mathbf{C}^e (\Delta\varepsilon - \Delta\varepsilon_0)$$

$$\Delta\varepsilon_0 = \Delta\varepsilon^P = \mathbf{L}^{-1} \left(\frac{\partial \mathbf{f}}{\partial \sigma} \right)^T \frac{\partial g(\sigma)}{\partial \sigma} \Delta\sigma \tag{5.84}$$

Then the equilibrium equation that the displacement increment $\Delta\mathbf{U}$ should satisfy is:

$$\mathbf{K}^e \Delta\mathbf{U} = \Delta\mathbf{R} + \overline{\mathbf{R}}(\Delta\sigma) \tag{5.85}$$

where: $\mathbf{K}^e$ = stiffness matrix in an elastic calculation and

$$\overline{\mathbf{R}}(\Delta\sigma) = \int_{VOL} \mathbf{B}^T \mathbf{C}^e \Delta\varepsilon^P dV = \int_{VOL} \mathbf{B}^T \mathbf{C}^e \mathbf{L}^{-1} \left(\frac{\partial \mathbf{f}}{\partial \sigma} \right)^T \frac{\partial g(\sigma)}{\partial \sigma} \Delta\sigma dV \tag{5.86}$$

is the equivalent nodal forces, or the correction load transformed from the initial strain $\Delta\varepsilon_0$.

Since the correction load is dependent on the stress increment $\Delta\sigma$ that itself is an undetermined variable, an iterative process is needed for solving equation (5.85). Thus, the method of initial incremental strain is a combination of incremental and iterative methods. The iterative equation for the *i*-th load increment is:

$$\mathbf{K}^e \Delta'\mathbf{U}^{(j)} = \Delta'\mathbf{R} + {}^j\overline{\mathbf{R}}^{(j-1)} \qquad (j = 0, 1, \cdots) \tag{5.87}$$

If the (*j*-1)-th approximation to $\Delta^i U^{(j-1)}$ is known, $\Delta^i \varepsilon^{(j-1)}$ and $\Delta^i \sigma^{(j-1)}$ can be obtained. Using Eq.(5.86), the correction load for the succeeding iteration, ${}^j\overline{R}^{(j-1)}$, is calculated. Then solving equation (5.87) yields $\Delta^i U^{(j)}$ which can be used to start the next iteration. Repeat the above iteration until the difference of stress increments in two consecutive iterative steps is less than a given tolerance.

5.2.3.4 Comparison Among the Three Methods

The incremental tangent stiffness method obtains an approximate solution by adjusting the stiffness matrix when each load increment is added. Therefore, a new stiffness matrix needs to be formed in each incremental step leading to a considerable amount of computation.

As to the incremental initial stress and strain methods, the stiffness matrix remains same when each load increment is added. Therefore, in each incremental and iterative step, only the second right-hand terms in equation (5.87) need to be calculated in an equivalent backward substitution. The stiffness matrix need to be calculated only one time at the beginning of the entire FEM calculation. Thus, the amount of computation is less than that associated with the incremental tangent stiffness method.

With the incremental initial stress or strain method, whenever a load increment is added, an iterative process has to be proceeded on the initial stress or strain. When a plastic zone is large, the iteration of these two methods converges very slowly.

5. 3 GEOMETRIC NON-LINEARITY

A non-linear geometric problem means a problem with a large displacement or strain. As we know, the stiffness matrix of an object is dependent upon its geometric position. In a non-linear geometric problem, the equilibrium equation must be described by the geometric position after deformation. Thus, the stiffness matrix of an object is a function of its geometric deformation, and can be expressed as:

$$\mathbf{KU} - \mathbf{R} = 0 \tag{5.88}$$

where: $\mathbf{U}$ = displacement column matrix
$\mathbf{R}$ = external force column matrix
$\mathbf{K}$ = stiffness matrix and

$$\mathbf{K} = \mathbf{K(U)}$$

There are three approaches inorder to solve the non-linear equations in equation (5.88). The first one is based upon the Newton-Raphson iterative method. For simplicity, consider a single freedom system and let

$$\Psi(\mathbf{U}) = \mathbf{K(U)U} - \mathbf{R} = 0 \tag{5.89a}$$

If the Newton-Raphson method is applied, we obtain the following iterative formula:

$$^{t}\mathbf{K}_{T}\Delta^{t+1}\mathbf{U} = \mathbf{R} - F(^{t}\mathbf{U})$$
$$^{t+1}\mathbf{U} = {}^{t}\mathbf{U} + \Delta^{t+1}\mathbf{U}$$

(5.89b)

where the superscripts to the right of the variables denote the number of iteration, and

$$^{t}\mathbf{K}_{T} = {}^{t}\left(\frac{d\Psi}{d\mathbf{U}}\right)$$

Fig. (5.12a) shows the process of using equation (5.89b) to solve equation (5.88). A modified Newton-Raphson method inidcates that the stiffness matrix does not change in the entire iterative process, as shown in Fig. (5.12b).

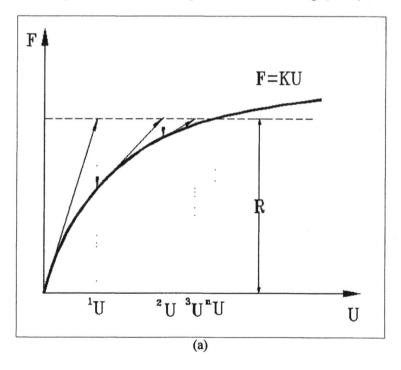

(a)

Figure 5.12 The Newton-Raphson method.

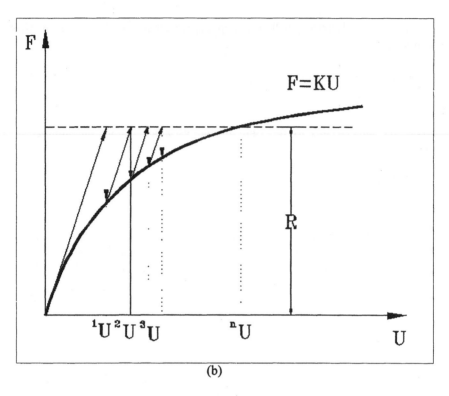

(b)

Figure 5.12   Continued.

The second approach is the incremental load method, as illustrated in Fig. 5.13. Its iterative method is:

$${}^{t}\mathbf{K}_{T}\Delta^{t+1}\mathbf{U} = \Delta^{t}\mathbf{R}$$
$${}^{t+1}\mathbf{U} = {}^{t}\mathbf{U} + \Delta^{t+1}\mathbf{U}$$

$$(5.90)$$

where: $\Delta^{t}\mathbf{R}$ = the i-th load increment

It is seen that the more finely the load is divided, the more an accurate prediction can be obtained.

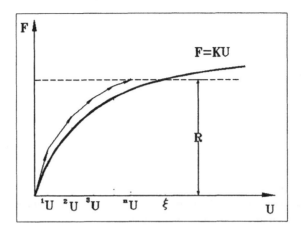

Figure 5.13 Incremental load method.

The third way is a combination of the previous two, i.e., the incremental and iterative method. In this method, the total load is divided into several subloads. During each load increment, a Newton-Raphson iteration or a modified one is applied.

In incremental non-linear finite element analysis, all variables can be referred to either an previous configuration or the initial configuration of the continuum, which corresponds to Eulerian and Total Lagrangian formulations, respectively. If these variables are referred to the recently computed configuration, the approach is called Updated Lagrangian or Approximate Eulerian. The Updated Lagrangian approach was found to be more general and computationally efficient (Desai and Phan, 1979),

Strain can be defined using either geometric linear or non-linear formulations as explained below. As an example, a two-dimensional four-node isoparametric element is considered in the following formulation.

5.3.1 Geometric Linear Strain Formula

According to the small-strain theory, a strain vector in two-dimensional space can be written as:

$$\varepsilon = \begin{bmatrix} \varepsilon_x \\ \varepsilon_y \\ \tau_{xy} \end{bmatrix} = \begin{bmatrix} u_{,x} \\ v_{,y} \\ u_{,y} + v_{,x} \end{bmatrix} \tag{5.91}$$

and

$$u_{,i} = \frac{\partial u}{\partial i}; \quad v_{,i} = \frac{\partial v}{\partial i} \quad (i = x, y)$$

$$\mathbf{u}_f = \sum_{i=1}^{4} N_i \mathbf{u}_i \tag{5.92}$$

$$\mathbf{u}_f = \begin{bmatrix} u \\ v \end{bmatrix}; \quad \mathbf{u}_i = \begin{bmatrix} u_i \\ v_i \end{bmatrix}$$

where: N_i = shape function
$\mathbf{u}_f$ = displacement column matrix at one point inside an element
$\mathbf{u}_i$ = nodal displacement column matrix

The strain at one point inside a four-node isoparametric element can be related to nodal displacement by the following equation:

$$\varepsilon = \mathbf{B}^L \mathbf{u} \tag{5.93}$$

and

$$\mathbf{B}^L = \begin{bmatrix} N_{1,x} & 0 & N_{2,x} & 0 & N_{3,x} & 0 & N_{4,x} & 0 \\ 0 & N_{1,y} & 0 & N_{2,y} & 0 & N_{3,y} & 0 & N_{4,y} \\ N_{1,y} & N_{1,x} & N_{2,y} & N_{2,x} & N_{3,y} & N_{3,x} & N_{4,y} & N_{4,x} \end{bmatrix} \tag{5.94}$$

$$\mathbf{u}^T = \begin{bmatrix} u_1 & v_1 & u_2 & v_2 & u_3 & v_3 & u_4 & v_4 \end{bmatrix}$$

where:

$$N_{i,x} = \frac{\partial N_i}{\partial x}; \quad N_{i,y} = \frac{\partial N_i}{\partial y} \tag{5.95}$$

The element stiffness matrix is then given by the following equation:

$$\mathbf{k}^L = \iint (\mathbf{B}^L)^T \mathbf{C} \mathbf{B}^L \, t \, dx \, dy \tag{5.96}$$

where: $\mathbf{B}^L$ = geometric linear strain matrix
$\mathbf{C}$ = constitutive matrix for soil elements and, in plane strain cases, is expressed by:

$$C = \frac{E(1-v)}{(1+v)(1-2v)} \begin{bmatrix} 1 & \dfrac{v}{1-v} & 0 \\ \dfrac{v}{1-v} & 1 & 0 \\ 0 & 0 & \dfrac{1-2v}{2(1-v)} \end{bmatrix} \tag{5.97}$$

Since the stress-stain relation of agricultural soils is usually in a non-linear form, incremental analysis should be used in the following form:

$$\mathbf{k}^L \Delta \mathbf{u} = \Delta \mathbf{r} \quad \text{and}$$
$$\mathbf{K}^L \Delta \mathbf{U} = \Delta \mathbf{R} \tag{5.98}$$

where: $\Delta \mathbf{r}$ = element incremental external load column matrix

$\Delta \mathbf{u}$ = element incremental displacement column matrix

$\Delta \mathbf{R}$ = global incremental external load column matrix

$\Delta \mathbf{U}$ = global incremental displacement column matrix

$\mathbf{k}^L$ = element stiffness matrix associated with $\mathbf{B}^L$

$\mathbf{K}^L$ = global stiffness matrix associated with $\mathbf{B}^L$

5.3.2 Geometric Non-linear Strain Formula

In the Updated Lagrangian approach (Bathe et al., 1975), a strain vector in two-dimensional space can be divided into linear and non-linear parts:

$$\varepsilon = \varepsilon^L + \varepsilon^{NL} \tag{5.99}$$

where:

$$\varepsilon^L = \begin{bmatrix} \varepsilon_x^L \\ \varepsilon_y^L \\ \tau_{xy}^L \end{bmatrix} = \begin{bmatrix} u_{,x} \\ v_{,y} \\ u_{,y} + v_{,x} \end{bmatrix}$$

$$\varepsilon^{NL} = \begin{bmatrix} \varepsilon_x^{NL} \\ \varepsilon_y^{NL} \\ \tau_{xy}^{NL} \end{bmatrix} = \begin{bmatrix} 0.5(u_{,x}^2 + v_{,x}^2) \\ 0.5(u_{,y}^2 + v_{,y}^2) \\ u_{,x}u_{,y} + v_{,x}v_{,y} \end{bmatrix} \tag{5.100}$$

The relation between the linear strain component ε^L and nodal displacement is given by

$$\varepsilon^L = \mathbf{B}^L \mathbf{u} \tag{5.101}$$

where $\mathbf{B}^L$ and $\mathbf{u}$ are determined by equation (5.94).

The incremental finite element formulation including geometric non-linearity can be written as (Bathe et al., 1975; Desai and Phan, 1979):

$$(\mathbf{k}^L + \mathbf{k}^{NL})\mathbf{u} = \Delta\mathbf{r} \quad \text{and}$$
$$(\mathbf{k}^L + \mathbf{k}^{NL})\mathbf{u} = \Delta\mathbf{R} \tag{5.102}$$

where $\mathbf{k}^L$ is determined by equation (5.96) and

$$\mathbf{k}^{NL} = \iint (\mathbf{B}^{NL})^T \mathbf{S} \mathbf{B}^{NL} \, t \, dx \, dy$$

$$\mathbf{B}^{NL} = \begin{bmatrix} N_{1,x} & 0 & N_{2,x} & 0 & N_{3,x} & 0 & N_{4,x} & 0 \\ N_{1,y} & 0 & N_{2,y} & 0 & N_{3,y} & 0 & N_{4,y} & 0 \\ 0 & N_{1,x} & 0 & N_{2,x} & 0 & N_{3,x} & 0 & N_{4,x} \\ 0 & N_{1,y} & 0 & N_{2,y} & 0 & N_{3,y} & 0 & N_{4,y} \end{bmatrix} \tag{5.103}$$

$$\mathbf{S} = \begin{bmatrix} \sigma_x & \tau_{xy} & 0 & 0 \\ \tau_{yx} & \sigma_y & 0 & 0 \\ 0 & 0 & \sigma_x & \tau_{xy} \\ 0 & 0 & \tau_{yx} & \sigma_y \end{bmatrix}$$

where: $\mathbf{B}^{NL}$ = geometric non-linear strain matrix

$\mathbf{k}^{NL}$ = element stiffness matrix associated with $\mathbf{B}^{NL}$

$\mathbf{S}$ = element stress matrix

5.3.3 Numerical Simulation of Large Soil Deformation

Large soil deformation and strain often occur in the soil-tool interaction. In order to evaluate the prediction accuracy of such deformation by different numerical approaches, finite element analysis of the deformation of soil specimens on a triaxial apparatus was conducted using different strain formulations and tangential stiffness methods, and the numerical results will be compared with experimental results below.

The idealized soil-specimen model is illustrated in Fig. 5.14. For simplicity, only plane strain analysis is conducted in this study with a thickness of 52 mm. The boundary conditions are given as follows:

1.   The nodes on the top boundary line (CD) are fixed in both horizontal and vertical directions.

2.   The nodes on the right-side and left-side boundary lines (AD and BC) are subject to a constant external load toward the center of soil specimen in horizontal direction.

3.   The nodes on the bottom boundary line (AB) have a specified upward displacement in vertical direction during each increment loading step and are fixed in horizontal direction.

4.   All other nodes are free in two directions.

The Duncan and Chang model is chosen as the soil constitutive model in this section. Since the tangential modulus in this model depends upon the soil stress state during a loading process, an incremental method is used to solve the soil non-linearity. The total specified displacement on line AB in Fig. 5.14 is divided into many small increments. During each increment, soil tangent modulus at each Gauss point is updated according to current stress status.

The general matrix differential equation in the FEM analysis is expressed by equations (5.98) and (5.102) corresponding to the geometric linear and non-linear strain formula, respectively. The updated Lagrangian method is adopted in all cases, that is, after each incremental step, a continuous node coordinate transformation is made.

The schematic view of a triaxial apparatus is shown in Fig. 5.15. The axial load to soil specimens is applied by an ELE 10 kN loading frame (model EL25-220) with a modified motor and transmission system which provided axial deformation rates from 0.002 mm/s to 12 mm/s (or axial strain rates from 0.000019 s to 0.113 s^{-1}). The confining pressure inside the test chamber is kept constant during each test by an pressure regulating system, and is measured by a standard pressure transducer (model PSI-100). The axial deformation of soil specimens is measured by a sealed super-mini load cell (model SSM-AS-500). The volume change inside soil specimens during a loading process can be measured by a pressure transducer on the basis of the measurement principle developed by Dunlap and Weber (1971). The typical calibration result shown in Fig. 5.16 indicates that Dunlap and Weber's method is accurate enough to measure the volume change inside an unsaturated soil specimen. All the above four signals related to confining pressure, axial deformation, axial load and volume change of soil specimen, are acquired by a personal computer through a Campbell Scientific 21X Micrologger.

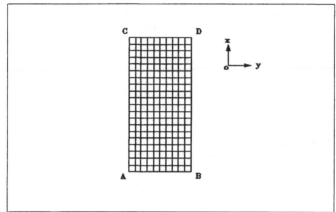

Figure 5.14 The finite element mesh.

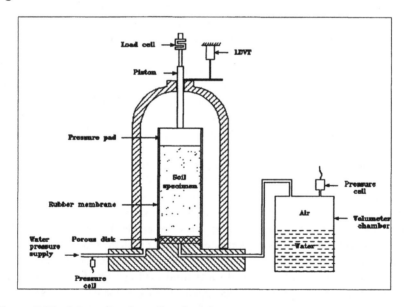

Figure 5.15 Schematic view of a triaxial apparatus system.

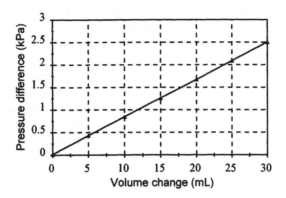

Figure 5.16 Calibration of the Dunlap-Weber volumeter.

5.3.3.1 Prediction Accuracy of Different Strain Formula

In order to evaluate the prediction accuracy of soil deformation, the average diameter of soil specimen is chosen as an index determined by the parameters that can be measured continuously in triaxial tests. This index is expressed by the following equation (Kezdi, 1980):

$$D_a = D_0 \sqrt{\frac{(1 - \varepsilon_v)}{(1 - \varepsilon)}} \tag{5.104}$$

where: D_a = average diameter of soil specimens after load
   D_0 = original diameter of soil specimens
   ε = axial strain of soil specimens on a triaxial apparatus
   ε_v = volumetric strain of soil specimens expressed by:

$$\varepsilon = \frac{\Delta h}{h_0}$$

$$\varepsilon_v = \frac{\Delta V}{V_0}, \quad positive \ when \ in \ compression$$

where: h_0 = original height of soil specimens
   Δh = axial deformation of soil specimens
   V_0 = original volume of soil specimens
   ΔV = volume change of soil specimens.

Fig. 5.17 shows the predicted average diameter by linear and non-linear geometric formulations in the incremental FEM analysis with the updated Lagrangian scheme. The difference between two approaches is very small. The reason for this is that a node coordinate transformation after each incremental step makes ε^{NL} in equation (5.100) and k^{NL} in equation (5.102) neglectable, and thus results in no noticeable difference in the FEM results using the two strain formula. Since the first approach has a simpler formulation and higher computation efficiency, compared to the second one, linear geometric formulation is suggested for incremental FEM analysis with the updated Lagrangian scheme.

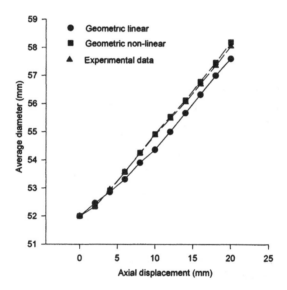

Figure 5.17 Predicted average diameter by linear and non-linear geometric formula.

5.3.3.2 Deformation Associated with Linear and Non-Linear Material Analysis

In an analysis of soil deformation, the material property can be considered either linear or non-linear. The linear material property means that the tangential modulus is a constant, whereas the non-linear denotes that the tangential modulus depends on stress and/or strain status as shown in equations (3.45) and

(3.46b). Since the Duncan-Chang model is chosen, the value of tangential modulus for linear material analysis is determined by the following expression:

$$E_0 = K_s p_a \left(\frac{\sigma_3}{p_a}\right)^n \tag{5.105}$$

where: E_0 = initial tangent modulus for soil elements
K and n = coefficients of Duncan-Chang model

p_a = atmospheric pressure

σ_3 = minor principal stress in soil

Fig. (5.18a) shows that the prediction of the average diameter of soil specimen is very close to each other for the two approaches (constant and variable tangential stiffness). However, the force prediction by these two approaches is very different, as shown in Fig. (5.18b). It is obvious that the force prediction by the variable tangential stiffness method is closer to the hyperbolic stress-strain relation reported by many researchers.

5.3.3.3 Influence of Poisson's Ratio on Deformation Prediction
According to Duncan and Chang's study (1970), Poisson's ratio can be calculated by the following incremental form:

$$v = \frac{\Delta\varepsilon - \Delta\varepsilon_v}{2\Delta\varepsilon} \tag{5.106}$$

where, $\Delta\varepsilon$ = increment of axial strain of soil specimens

$\Delta\varepsilon_v$ = increment of volume strain of soil specimens

Fig. 5.19 illustrates the variation of Poisson's ratio obtained from triaxial tests under different confining pressures. It can be seen from this figure that experimental Poisson's ratio continuously changed during a loading process. However, for simplicity, a single constant Poisson's ratio is used to represent its average value during each loading process. In this study, the single average of Poisson's ratio is chosen as 0.4 for all four loading cases.

The influence of Poisson's ratio on deformation prediction is shown in Fig. 5.20. Compared to the influence by different strain formulations, Poisson's ratio appears to be more important. Therefore, the accurate measurement of Poisson's ratio is a key point for the FEM analysis of soil deformation, such as soil compaction prediction.

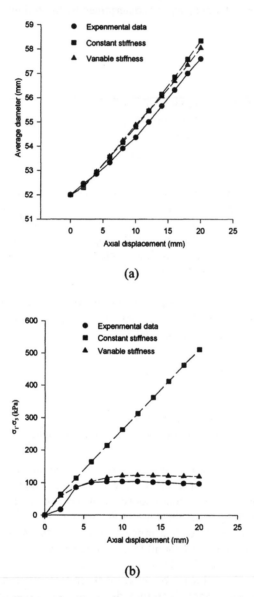

Figure 5.18 Prediction of average diameter by constant and variable tangential stiffness approaches.

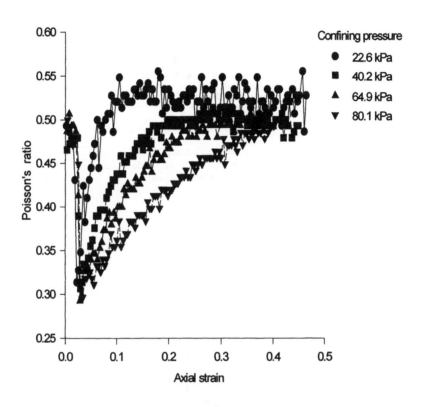

Figure 5.19 Poisson's ratio obtained from triaxial tests.

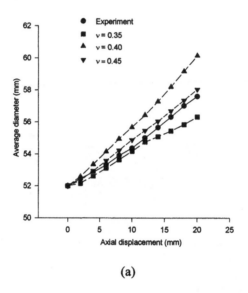

(a)

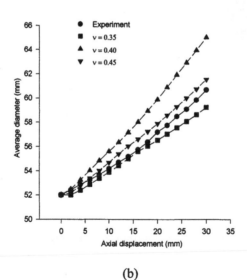

(b)

Figure 5.20 Prediction of average diameter with different Poisson's ratio.
(a)$\sigma_3 = 22.6$ kPa; (b) $\sigma_3 = 40.2$ kPa; (c) $\sigma_3 = 64.9$ kPa; (d) $\sigma_3 = 80.1$ kPa.

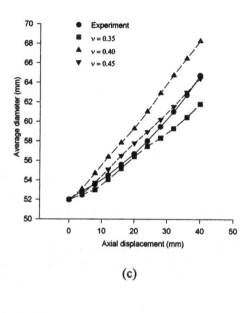

(c)

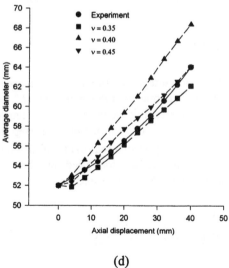

(d)

Figure 5.20 Continued.

5.4   DYNAMIC RESPONSE

Problems in soil-tool systems are usually dynamic. The dynamic response of
soil can be simulated by one of four direct integration methods: the central
difference method, the Houbolt method, the Newmark method, and the Wilson-
θ method. The main weakness of the central difference method is that its
solution is only conditionally stable, i.e., the time step must be smaller than a
critical time step to obtain a stable solution. The inconvenience associated to the
Houbolt method is the requirement of a special starting procedure to obtain the
displacement at two instants Δt and $2\Delta t$, where Δt is the time interval. Both
the Newmark method and the Wilson-θ method belong to the linear
acceleration method, and are unconditionally stable for linear cases if
$\gamma \geq 0.5$; $\beta \geq 0.25(0.5 + \gamma)^2$ for the Newmark method and $\theta \geq 1.37$ for the
Wilson-θ method. In this section, only the algorithm implementation based on
the Newmark method is discussed, but a similar procedure can be easily applied
to the Wilson-θ method.

The conventional Newmark step-by-step integration method is not well suited
in solving a problem with the coupling of dynamic response and material non-
linearity. In this section, an incremental and iterative algorithm for dynamic
soil-machine interactions is proposed.

The dynamic equation for a discrete finite element system can be expressed
as:

$$\mathbf{M}\ddot{\mathbf{U}} + \mathbf{C}\dot{\mathbf{U}} + \mathbf{K}\mathbf{U} = \mathbf{R} \tag{5.107}$$

The main idea of step-by-step integration methods is to loosen the requirement
for "satisfaction of the basic dynamic equation at any time t " to the requirement
for "satisfaction of the basic dynamic equation at only limited discrete time
points: $0, \Delta t, 2\Delta t, 3\Delta t, \cdots$", where Δt is an incremental step in time chosen to
coincide with the incremental step in external loads. If all kinematic and static
variables at different time points are identified by their left superscripts,
equation (5.107) at time t and $t + \Delta t$ is expressed as:

$$
\begin{aligned}
&{}^{t}\mathbf{M}\,{}^{t}\ddot{\mathbf{U}} + {}^{t}\mathbf{C}\,{}^{t}\dot{\mathbf{U}} + {}^{t}\mathbf{K}\,{}^{t}\mathbf{U} = {}^{t}\mathbf{R}; \\
&{}^{t+\Delta t}\mathbf{M}\,{}^{t+\Delta t}\ddot{\mathbf{U}} + {}^{t+\Delta t}\mathbf{C}\,{}^{t+\Delta t}\dot{\mathbf{U}} + {}^{t+\Delta t}\mathbf{K}\,{}^{t+\Delta t}\mathbf{U} = {}^{t+\Delta t}\mathbf{R}
\end{aligned}
\tag{5.108}
$$

Assuming ${}^{t+\Delta t}\mathbf{M} \approx {}^{t}\mathbf{M}$, ${}^{t+\Delta t}\mathbf{C} \approx {}^{t}\mathbf{C}$, ${}^{t+\Delta t}\mathbf{K} \approx {}^{t}\mathbf{K}$, if the incremental step is small
enough, the subtraction of the first equation from the second equation in

equation (5.108) yields the following basic dynamic equation in increment form:

$$^t\mathbf{M}\Delta^{t+\Delta t}\ddot{\mathbf{U}}+{}^t\mathbf{C}\Delta^{t+\Delta t}\dot{\mathbf{U}}+{}^t\mathbf{K}\Delta^{t+\Delta t}\mathbf{U}=\Delta^{t+\Delta t}\mathbf{R} \tag{5.109}$$

where:

$$\begin{aligned}
\Delta^{t+\Delta t}\mathbf{U}&=^{t+\Delta t}\mathbf{U}-{}^t\mathbf{U}\\
\Delta^{t+\Delta t}\dot{\mathbf{U}}&=^{t+\Delta t}\dot{\mathbf{U}}-{}^t\dot{\mathbf{U}}\\
\Delta^{t+\Delta t}\ddot{\mathbf{U}}&=^{t+\Delta t}\ddot{\mathbf{U}}-{}^t\ddot{\mathbf{U}}\\
\Delta^{t+\Delta t}\mathbf{R}&=^{t+\Delta t}\mathbf{R}-{}^t\mathbf{R}
\end{aligned} \tag{5.110}$$

5.4.1 Algorithm Associated with the Newmark Method

Formulations
According to the Lagrangian median theorem,

$$^{t+\Delta t}\dot{\mathbf{U}}={}^t\dot{\mathbf{U}}+\tilde{\ddot{\mathbf{U}}}\Delta t \tag{5.111}$$

where $\tilde{\ddot{\mathbf{U}}}$ is value of $\ddot{\mathbf{U}}$ at certain point in the domain $[t,\,t+\Delta t]$. The Newmark method assumes:

$$\tilde{\ddot{\mathbf{U}}}=(1-\gamma)\,^t\ddot{\mathbf{U}}+\gamma\,^{t+\Delta t}\ddot{\mathbf{U}}\qquad 0\le\gamma\le1 \tag{5.112}$$

Therefore, equation (5.111) is rewritten as:

$$\Delta^{t+\Delta t}\dot{\mathbf{U}}=\Delta t\,^t\ddot{\mathbf{U}}+a_0\,\Delta^{t+\Delta t}\ddot{\mathbf{U}} \tag{5.113}$$

where: $a_0=\gamma\,\Delta t$

According to the Taylor expansion of the displacement vector,

$$^{t+\Delta t}\mathbf{U}={}^t\mathbf{U}+{}^t\dot{\mathbf{U}}\Delta t+(0.5\Delta t^2)\,^{t+h\Delta t}\ddot{\mathbf{U}}\qquad 0<h<1 \tag{5.114}$$

The Newmark method assumes:

$$^{t+h\Delta t}\ddot{U} = (1 - 2\beta)\,{}^{t}\ddot{U} + 2\beta\,{}^{t+\Delta t}\ddot{U} \qquad 0 \le 2\beta \le 1 \tag{5.115}$$

Thus, equation (5.114) can be rewritten as:

$$\Delta^{t+\Delta t}\ddot{U} = a_1\,\Delta^{t+\Delta t}U + a_2\,{}^{t}\dot{U} + a_3\,{}^{t}\ddot{U} \tag{5.116}$$

where:

$$a_1 = \frac{1}{\beta\Delta t^2}; \quad a_2 = -\frac{1}{\beta\Delta t}; \quad a_3 = -\frac{1}{2\beta}$$

By the substitution of equations (5.113) and (5.116) into equation (5.109) and after several rearrangements, equation (5.109) is rewritten as:

$$(a_1\,{}^{t}M + a_4\,{}^{t}C + {}^{t}K)\,\Delta^{t+\Delta t}U$$
$$= \Delta^{t+\Delta t}R + {}^{t}M(-a_2\,{}^{t}\dot{U} - a_3\,{}^{t}\ddot{U}) + {}^{t}C(-a_5\,{}^{t}\dot{U} - a_6\,{}^{t}\ddot{U}) \tag{5.117}$$

where: $a_4 = a_0 a_1; \quad a_5 = a_0 a_2; \quad a_6 = \Delta t + a_0 a_3$

The unknown displacement column matrix, $\Delta^{t+\Delta t}U$, can be solved, if ${}^{t}U$, ${}^{t}\dot{U}$ and ${}^{t}\ddot{U}$ are known. Then, on the basis of equations (5.113) and (5.116), ${}^{t+\Delta t}U$, ${}^{t+\Delta t}\dot{U}$ and ${}^{t+\Delta t}\ddot{U}$ are determined by the following equations:

$$^{t+\Delta t}\ddot{U} = a_1\,\Delta^{t+\Delta t}U + a_2\,{}^{t}\dot{U} + a_7\,{}^{t}\ddot{U}$$
$$^{t+\Delta t}\dot{U} = {}^{t}\dot{U} + a_8\,{}^{t}\ddot{U} + a_0\,{}^{t+\Delta t}\ddot{U}$$
$$^{t+\Delta t}U = {}^{t}U + \Delta^{t+\Delta t}U \tag{5.118}$$
$$a_7 = a_3 + 1; \quad a_8 = \Delta t - a_0$$

Procedure
The procedure of implementing the Newmark scheme in an incremental form along with the Newton-Raphson iteration is explained below in a FORTRAN-like algorithmic language in which all bold strings belong to FORTRAN keywords. The left and right superscripts refer to incremental and iterative steps, respectively.

1: Input control information:

        NSTEP ---- number of total incremental steps

      NITEM ---- number of maximum iteration steps in each incremental
         step

      γ and β ---- $\gamma \geq 0.5$; $\beta \geq 0.25(0.5 + \gamma)^2$

      tol ---- error tolerance

2:   Calculate constants $a_0 - a_8$ based on equations (5.113), (5.116), (5.117) and (5.118)

3:   Initialize the displacement, velocity and acceleration column matrices:

$$^{0}\mathbf{U},\ ^{0}\dot{\mathbf{U}}\ \text{and}\ ^{0}\ddot{\mathbf{U}}$$

4:   **DO** 23 i=1, *NSTEP*

5:   Form mass matrix $^{i-1}\mathbf{M}$ and damping matrix $^{i-1}\mathbf{C}$ at increment step i-1

6:   Form stiffness matrix $^{i-1}\mathbf{K}$ and equivalent stiffness matrix $^{i-1}\tilde{\mathbf{K}}$ at increment step i-1

$$^{i-1}\tilde{\mathbf{K}} = a_1\ ^{i-1}\mathbf{M} + a_4\ ^{i-1}\mathbf{C} + ^{i-1}\mathbf{K}$$

7:   Zero the following column matrices:

      $\Delta^{i}\mathbf{U}$ ---- increment of the displacement column matrix $^{i}\mathbf{U}$ at increment
         step i

      $\Delta^{i}\mathbf{R}$ ---- increment of the load column matrix $^{i}\mathbf{R}$ at increment step i

8:   Calculate $\Delta^{i}\mathbf{R}$ on the basis of external loads and specified displacements

9:   Form effective load column matrix $\Delta^{i}\tilde{\mathbf{R}}$:

$$\Delta^{i}\tilde{\mathbf{R}} = \Delta^{i}\mathbf{R} + ^{i-1}\mathbf{M}(-a_2\ ^{i-1}\dot{\mathbf{U}} - a_3\ ^{i-1}\ddot{\mathbf{U}}) + ^{i-1}\mathbf{C}(-a_5\ ^{i-1}\dot{\mathbf{U}} - a_6\ ^{i-1}\ddot{\mathbf{U}})$$

10: Solve for displacement increment column matrix $\Delta^{i}\mathbf{U}$:

$$^{i-1}\tilde{\mathbf{K}}\Delta^{i}\mathbf{U} = \Delta^{i}\tilde{\mathbf{R}}$$

11: $\Delta^{i}\mathbf{U}^{(0)} = \Delta^{i}\mathbf{U}$

12: **DO** 20 ii=1,*NITEM*

13: Zero the following column matrices:

      $\Delta(\Delta^{i}\mathbf{U})^{(ii)}$ ---- increment of $\Delta^{i}\mathbf{U}$ in ii-th iteration

$\Delta(\Delta'\mathbf{R})^{(n)}$ ---- increment of $\Delta'\mathbf{R}$ in ii-th iteration

14: Calculate (ii-1)st approximation to acceleration column matrix $\Delta'\ddot{\mathbf{U}}^{(ii-1)}$ and velocity column matrix $\Delta'\dot{\mathbf{U}}^{(n-1)}$:

$$\Delta'\ddot{\mathbf{U}}^{(ii-1)} = a_1 \Delta'\mathbf{U}^{(ii-1)} + a_2 {}^{t-1}\dot{\mathbf{U}} + a_3 {}^{t-1}\ddot{\mathbf{U}}$$
$$\Delta'\dot{\mathbf{U}}^{(ii-1)} = \Delta t {}^{t-1}\ddot{\mathbf{U}} + a_0 \Delta'\ddot{\mathbf{U}}^{(ii-1)}$$

15: Calculate *(ii-1)*th approximation to strain and stress increment column matrices:

$${}^t\Delta\varepsilon^{(n-1)} \ and \ {}^t\Delta\sigma^{(n-1)}$$

16: Calculate the increment of the internal force column matrix $\Delta'\mathbf{F}^{(ii-1)}$

$$\Delta'\mathbf{F}^{(n-1)} = \int({}^t\mathbf{B}^T)^{(ii-1)} \ {}^t\Delta\sigma^{(n-1)}dV$$

17: Calculate the increment of effective out-of-balance load vector in ii-th iteration:

$$\Delta(\Delta'\mathbf{R})^{(n)} = \Delta'\mathbf{R} - \Delta'\mathbf{F}^{(ii-1)} + {}^{t-1}\mathbf{M}\Delta'\ddot{\mathbf{U}}^{(ii-1)} + {}^{t-1}\mathbf{C}\Delta'\dot{\mathbf{U}}^{(n-1)}$$

18: Solve for ii-th correction to displacement increment column matrix:

$${}^{t-1}\widetilde{\mathbf{K}}\Delta(\Delta'\mathbf{U})^{(ii)} = \Delta(\Delta'\mathbf{R})^{(n)}$$

19: Calculate new displacement increment column matrix at ii-th iteration:

$$\Delta'\mathbf{U}^{(n)} = \Delta'\mathbf{U}^{(n-1)} + \Delta(\Delta'\mathbf{U})^{(n)}$$
$$\Delta'\mathbf{U} = \Delta'\mathbf{U}^{(ii)}$$

20: **IF**$(\|\Delta(\Delta'\mathbf{U})^{(ii)}\|_2 / \|\Delta'\mathbf{U}^{(n)}\|_2 . LT. tol)$ **GOTO** 21
21: **Continue**
22: Calculate new displacement, velocity and acceleration column matrices in increment step i:

$$'\ddot{U} = a_1 \Delta'U + a_2 \,{}^{i-1}\dot{U} + a_7 \,{}^{i-1}\ddot{U}$$

$$'\dot{U} = {}^{i-1}\dot{U} + a_8 \,{}^{i-1}\ddot{U} + a_0 \,'\ddot{U}$$

$$'U = {}^{i-1}U + \Delta^i U$$

23: Update coordinates of nodal point, stress, strain and tangential modulus in
   each element
24: **Continue**
25: **End**

5.4.2 Algorithm Associated with the Wilson-θ Method

Formulations
The main difference between the Newmark method and Wilson-θ method is
that the latter assumes linear change of acceleration in time range [t, t+$\theta\Delta t$].
Therefore, the acceleration at time t+τ can be written as:

$$^{t+\tau}\ddot{U} = {}^{t}\ddot{U} + \frac{\tau}{\theta\Delta t}\left({}^{t+\theta\Delta t}\ddot{U} - {}^{t}\ddot{U}\right) \tag{5.119}$$

Integration of the above equation yields:

$$^{t+\tau}\dot{U} = {}^{t}\dot{U} + \tau\,{}^{t}\ddot{U} + \frac{\tau^2}{2\theta\Delta t}\left({}^{t+\theta\Delta t}\ddot{U} - {}^{t}\ddot{U}\right) \tag{5.120}$$

$$^{t+\tau}U = {}^{t}U + \tau\,{}^{t}\dot{U} + \frac{\tau^2}{2}\,{}^{t}\ddot{U} + \frac{\tau^3}{6\theta\Delta t}\left({}^{t+\theta\Delta t}\ddot{U} - {}^{t}\ddot{U}\right) \tag{5.121}$$

Let $\tau = \theta\Delta t$ and substitute it into the above two equations and rearrange these
equations leading to:

$$^{t+\Delta t}\ddot{U} = a_1 \Delta^{t+\theta\Delta t}U + a_2\,'\dot{U} + a_3\,'\ddot{U}$$

$$^{t+\Delta t}\dot{U} = a_4 \Delta^{t+\theta\Delta t}U + a_5\,'\dot{U} + a_6\,'\ddot{U} \tag{5.122}$$

where:

$$a_1 = \frac{6}{\theta^2 \Delta t^2}; \quad a_2 = -\frac{6}{\theta \Delta t}; \quad a_3 = -3$$

$$a_4 = \frac{3}{\theta \Delta t}; \quad a_5 = a_3; \quad a_6 = -\frac{\theta \Delta t}{2}$$

(5.123)

and

$$\Delta^{t+\theta \Delta t} \mathbf{U} = {}^{t+\theta \Delta t} \mathbf{U} - {}^{t} \mathbf{U}$$

$$\Delta^{t+\theta \Delta t} \dot{\mathbf{U}} = {}^{t+\theta \Delta t} \dot{\mathbf{U}} - {}^{t} \dot{\mathbf{U}}$$

$$\Delta^{t+\theta \Delta t} \ddot{\mathbf{U}} = {}^{t+\theta \Delta t} \ddot{\mathbf{U}} - {}^{t} \ddot{\mathbf{U}}$$

(5.124)

Assuming $^{t+\theta \Delta t} \mathbf{M} \cong {}^{t} \mathbf{M}$, $^{t+\theta \Delta t} \mathbf{C} \cong {}^{t} \mathbf{C}$, $^{t+\theta \Delta t} \mathbf{K} \cong {}^{t} \mathbf{K}$, the incremental dynamic equation in time range $[t, t+\theta \Delta t]$ has the following form similar to equation (5.109):

$${}^{t} \mathbf{M} \Delta^{t+\theta \Delta t} \ddot{\mathbf{U}} + {}^{t} \mathbf{C} \Delta^{t+\theta \Delta t} \dot{\mathbf{U}} + {}^{t} \mathbf{K} \Delta^{t+\theta \Delta t} \mathbf{U} = \Delta^{t+\theta \Delta t} \mathbf{R}$$

(5.125)

where:

$$\Delta^{t+\theta \Delta t} \mathbf{R} = {}^{t+\theta \Delta t} \mathbf{R} - {}^{t} \mathbf{R}$$

(5.126)

By the substitution of equation (5.123) into equation (5.125) and several rearrangements, equation (5.109) is transformed to:

$$(a_1 \, {}^{t} \mathbf{M} + a_4 \, {}^{t} \mathbf{C} + {}^{t} \mathbf{K}) \Delta^{t+\theta \Delta t} \mathbf{U}$$

$$= \Delta^{t+\theta \Delta t} \mathbf{R} + {}^{t} \mathbf{M}(-a_2 \, {}^{t} \dot{\mathbf{U}} - a_3 \, {}^{t} \ddot{\mathbf{U}}) + {}^{t} \mathbf{C}(-a_5 \, {}^{t} \dot{\mathbf{U}} - a_6 \, {}^{t} \ddot{\mathbf{U}})$$

(5.127)

The unknown displacement column matrix, $\Delta^{t+\theta \Delta t} \mathbf{U}$, can be solved, if ${}^{t} \mathbf{U}$, ${}^{t} \dot{\mathbf{U}}$ and ${}^{t} \ddot{\mathbf{U}}$ are known. Then, $^{t+\Delta t} \mathbf{U}$, $^{t+\Delta t} \dot{\mathbf{U}}$ and $^{t+\Delta t} \ddot{\mathbf{U}}$ are determined by the following equations:

$${}^{t+\Delta t} \ddot{\mathbf{U}} = a_0 \Delta^{t+\theta \Delta t} \mathbf{U} + a_7 \, {}^{t} \dot{\mathbf{U}} + a_8 \, {}^{t} \ddot{\mathbf{U}}$$

$${}^{t+\Delta t} \dot{\mathbf{U}} = {}^{t} \dot{\mathbf{U}} + 0.5 \Delta t \left({}^{t+\Delta t} \ddot{\mathbf{U}} + {}^{t} \ddot{\mathbf{U}} \right)$$

$${}^{t+\Delta t} \mathbf{U} = {}^{t} \mathbf{U} + \Delta t \, {}^{t} \dot{\mathbf{U}} + \frac{\Delta t^2}{6} \left({}^{t+\Delta t} \ddot{\mathbf{U}} + 2 \, {}^{t} \ddot{\mathbf{U}} \right)$$

(5.128)

where:

$$a_0 = \frac{6}{\theta^3 \Delta t^2}; \quad a_7 = \frac{6}{\theta^2 \Delta t}; \quad a_8 = 1 - \frac{3}{\theta} \tag{5.129}$$

Procedure

The procedure of implementing the Wilson-θ scheme in an incremental form along with the Newton-Raphson iteration is the same as that for the Newmark method except for the following differences:

1: Input control information:

   NSTEP ---- number of total incremental steps

   NITEM ---- number of maximum iteration steps in each incremental step

   θ ---- $\theta \geq 1.37$

   tol ---- error tolerance

2: Calculate constants $a_0 - a_8$ based on equations (5.123) and (5.129)

22: Calculate new displacement, velocity and acceleration column matrices in increment step i:

$$'\ddot{U} = a_0 \Delta' U + a_7 \, {}^{i-1}\dot{U} + a_8 \, {}^{i-1}\ddot{U}\}$$

$$'\dot{U} = {}^{i-1}\dot{U} + 0.5\Delta t\left('\ddot{U} - {}^{i-1}\ddot{U}\right)$$

$$'U = {}^{i-1}U + \Delta t \, {}^{i-1}\dot{U} + \frac{\Delta t^2}{6}\left('\ddot{U} + 2 \, {}^{i-1}\ddot{U}\right)$$

6

EXAMPLES OF SOIL-TOOL INTERACTION

Tillage and planting operations manipulate soil by mechanical action of soil engaging tools. Even though the operating depth seldom exceeds 100 mm, the quantity of soil moved each year by these tools is enormous, because of the large cultivation area is involved. Consequently, the energy used to move this soil is very large and accounts for nearly one-half of the total energy used in crop production system.

Much work has been reported on the static or dynamic analysis of soil-tool interaction in tillage operations using the FEM. In the following, we will present the most distinguished studies accomplished by investigators from different countries.

6.1 MILESTONES IN FINITE ELEMENT ANALYSIS OF SOIL-TOOL INTERACTION

Most investigators used a blade as the object studying the interaction between soil and tool, because its geometric simplicity made the corresponding FEM analysis relatively easy.

6.1.1 Plane Soil Cutting (Yong and Hanna)

Yong and Hanna (1977) used the finite element method as a theoretical means for determining soil behavior under the actions of a plane cutting blade. The soil mass is divided into many small elements with each element only connected to its adjacent elements at its nodal points. Fig. 6.1 shows a typical abstraction of a soil-blade system using the finite element mesh.

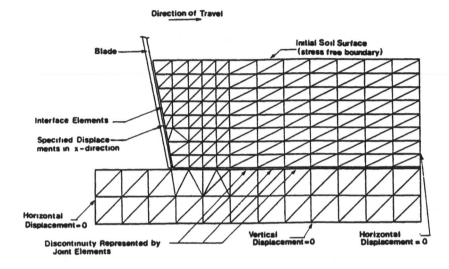

Figure 6.1 Abstraction of a blade-soil system using the finite element method. (From *Journal of Terramechanics*, R.N. Yong et. al., 1977. Reproduction with permission of The International Society for Terrain Vehicle Systems, Hanover, NH)

Three different types of elements are used in Fig. 6.1. The first one is the plane strain continuum element used for describing soil mass. However, Yong and Hanna considered that this type of element cannot satisfactorily model the soil deformation behavior in the situation where discontinuities may develop. They used the joint element proposed by Goodman et al. to describe the discontinuities at the soil-tool interface as well as a predefined horizontal slip surface right below the tool tip. It was emphasized that the predefined slip surface was essential to represent large relative displacements and separation of soil blocks right below the tool tip if an investigation was to be made on a continuous soil deformation process.

Constitutive Models for Soil and Interface
The stress-strain relation for soil mass elements is obtained from laboratory plane-strain tests with the incremental technique. Poisson's ratio is kept constant as 0.49 in the analysis. For the elements at the interface, shear stress-relative displacement curves at different normal stresses is obtained from direct shear

tests. The properties of interface elements consist of normal stiffness, K_n, and shear stiffness, K_s, which are related with stress and displacement as follows:

$$K_n s_n = \sigma_n$$
$$K_s s_s = \tau$$

(6.1)

where, σ_n and τ are normal and shear stresses, respectively; s_n and s_s are average relative normal and shear displacement within the element. The relation between τ and s_s is described by a hyperbolae of the following form:

$$\tau = \frac{s_s}{a + bs_s}$$

(6.2)

where a and b are empirical coefficients determined from direct shear tests. The tangent stiffness, K_{st}, denoting the slope of a shear stress-displacement curve, is expressed as:

$$K_{st} = \frac{1}{a}(1 - \tau b)^2$$

(6.3)

When an interface element is in compression state, K_n is assigned the value of initial elastic modulus of soil mass elements, and K_s is made equal to K_{st} in equation (6.3); After this element has failed, K_s is assigned a very small magnitude while K_n remains unchanged.

Geometric Non-linearity
The geometric non-linearity of soil deformation is tackled by updating the finite element mesh after each incremental step. In this way, a small displacement infinitesimal strain analysis is valid even for the case of very large displacement. The finite element mesh is modified by updating nodal coordinates with the following equation:

$$'\bar{x} = 'x + 'u \qquad '\bar{y} = 'y + 'v$$

(6.4)

where: $'\bar{x}$ and $'\bar{y}$ = updated nodal coordinates after the i-th incremental step

$'x$ and $'y$ = nodal coordinates at the beginning of the i-th incremental step

$'u$ and $'v$ = displacements in directions x and y in the i-th incremental step.

Experimental and Analytical Results

In the finite element analysis, the blades are considered as rigid. Uniform displacement increments are applied on all nodes at the interface. The total blade displacement is 2.5 cm that is divided into ten equal increments.

The results of the FEM analysis and corresponding experiments can be summarized in four aspects:

(1) Comparison of horizontal and vertical displacement fields obtained from experiment and numerical analysis.
(2) Horizontal, vertical and shear soil stress contour obtained from numerical analysis.
(3) Distribution of tangential and normal stresses as well as tangential displacements at interface obtained from numerical analysis.
(4) Comparison of measured and predicted tool drafts.

For the first aspect, agreement between two has been obtained, and a close agreement is seen in the fourth aspect.

One unsatisfactory point in the model is that the predefined horizontal slip surface using the joint element severely limits the usefulness of the model. The major advantage of the FEM over conventional analytical methods is that it does not require a predefined failure surface. In addition, the model may be suited only for the situation where the tool moves at a constant elevation.

6.1.2. Two-Dimensional Soil Cutting (Xie)

Following the work done by Yong and Hanna, Xie (1983) also conducted two-dimensional FEM analysis of soil cutting problem with some new considerations and experiments.

Soil and Interface Elements

For soil mass elements, the Duncan-Chang model (1970) is adopted to describe the soil stress-strain relation. The detailed expressions about this constitutive relation and the tangent modulus are given in Section 3.2.3.

As to interface elements, Xie considered that the joint element had some disadvantages. For instance, it can not reflect a certain proportional relation between normal and shear stresses during a slip at the interface. Besides, in a compression state, elements might penetrate each other, which makes the calculation results conflicting. A layer element is proposed with the following form of element stiffness matrix:

$$\mathbf{k} = \begin{bmatrix} \mathbf{k}_{ii} & \mathbf{k}_{ij} \\ \mathbf{k}_{ji} & \mathbf{k}_{jj} \end{bmatrix}$$ (6.5)

where:

$$\mathbf{k}_{ii} = \begin{cases} \begin{bmatrix} \alpha\cos^2\theta + \beta\sin^2\theta & (\alpha - \beta)\cos\theta\sin\theta \\ (\alpha - \beta)\cos\theta\sin\theta & \alpha\sin^2\theta + \beta\cos^2\theta \end{bmatrix} & \text{element in stick} \\[2em] \begin{bmatrix} \cos^2\theta \pm f\cos\theta\sin\theta & \theta\cos\theta\sin\theta \pm f\sin^2\theta \\ \cos\theta\sin\theta \mp f\cos^2\theta & \sin^2\theta \mp f\cos\theta\sin\theta \end{bmatrix} & \text{element in slip} \end{cases}$$ (6.6)

and

$$\mathbf{k}_{jj} = \mathbf{k}_{ii} \quad \mathbf{k}_{ij} = \mathbf{k}_{ji} = -\mathbf{k}_{ii}$$ (6.7)

where: α and β = parameters dependent upon material properties and geometry of the element
   f = friction coefficient at the interface

When the direction of relative slip is different, the plus and minus signs in equation (6.6) should be adjusted. At the beginning of an analysis, the layer interface element is assumed to be in an elastic state. After each incremental load is applied, first identify whether the element is in a stick or slip state, then make a corresponding adjustment based upon equation (6.6).

Experimental and Analytical Results
Normal and tangential stress sensors are embedded in the cutting blade. Comparison of measured and calculated distribution of tangential and normal stresses along the soil-tool interface indicates that the prediction of normal stress is very close to the experimental result, while that of tangential stress is not.

Besides measuring the blade draft, the vertical resistance force of blades was also acquired. Comparison of measured and calculated forces indicates that the prediction of total tool resistance is quite accurate, while that of the vertical resistance forces is not.

Possible soil failure zone is analyzed using an index, shear ratio S, which is defined as follows:

$$S = \frac{\sigma_1 - \sigma_3}{(\sigma_1 - \sigma_3)_f} = \frac{(\sigma_1 - \sigma_3)(2c\cos\phi + 2\sigma_3\sin\phi)}{1 - \sin\phi} \qquad (6.8)$$

which represents the ratio of the diameter of the actual Mohr stress circle to the diameter of the Mohr stress circle at failure and at the same σ_3. The value of S ranges from 0 to 1 representing soil from being not subject to shear to failed. The analyses indicate that the predicated failure zone is close to the measured one, and the smaller the rake angle, the better the prediction accuracy.

Tangent modulus, E_t, is used as an index to investigate the formation of soil wedge. The contours of tangent modulus of soil mass elements after the second incremental step, is investigated. Based upon different rake angles, the results are as follows:

(a) when rake angle $\alpha = 90°$ or $80° > 90°-\phi$, where $\phi = 25.6°$, there exists a densely-distributed E_t contour zone in front of the blades. This zone indicates that a compacted soil wedge will be formed and the volume of the wedge increases with α..

(b) when $\alpha = 30°$ or $45° < 90°-\phi$, the distribution of E_t contours is not dense, which indicates that no soil wedge will be generated.

(c) when $\alpha = 60° \cong 90°-\phi$, the distribution density of E_t contours is between the above two cases. In this critical case, a soil wedge may or may not be formed. In the experiment, no soil wedge was observed.

The above FEM prediction about the formation of soil wedge is consistent with the investigation by Tanner (1960). Furthermore, major and minor principal soil stress contours are given from numerical analysis.

6.1.3 Fuzzy and Dynamic Finite Element Analysis (Xie and Zhang)

Fuzzy Finite Element for Soil-Tool Semi-Infinite System

Xie and Zhang (1985) constructed a three-dimensional fuzzy element to simulate soil-tool semi-infinite system. The purpose of this study is to reduce the total number of degrees of freedom (d.o.f.) of the system without sacrificing the accuracy of the solution, leading to the cut down of computation time and storage space costs.

Theoretically, a soil-tool system should be defined on a semi-infinite domain. Conventional finite element method idealizes this semi-infinite domain to an enough large domain and then gives the corresponding boundary conditions. It is assumed that the influence from outside the specified domain is negligible. But this is true only for an enough large domain. For three dimensional analysis, in some cases computation time cost may become intolerable. Therefore, an effective way to dramatically reduce the total number of d.o.f. is of importance.

In order to simulate an infinite or semi-infinite domain problem, the entire system domain Ω may be divided into two subdomains R_1 and R_2 using a closed curve C, as illustrated in Fig. 6.2. In the figure, R_1 should be large enough to enclose a key subdomain R_0 of interest, while R_2 denotes the rest portion of the system domain whose boundary extends to infinite distance. Within R_1, conventional finite elements can be used, while in R_2 fuzzy elements might be implemented.

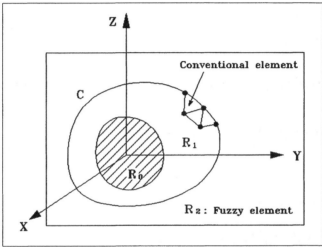

Figure 6.2 Division of a system domain. (From *Transactions of Chinese Society of Agricultural Machinery*, Xie. et al., 1985. Reproduced with permission of Chinese Society of Agricultural Machinery, Beijing, China.)

For a conventional finite element, its definition domain, *A*, is a clear set of Ω with the following relation:

$$A = \chi_A(u)\Omega \tag{6.9}$$

where: χ_A = a characteristic function of a set A
    u = an element

The stiffness matrix of this element can be expressed on the system domain as follows:

$$\mathbf{k} = \int_A \mathbf{B}^T \mathbf{CB} dV = \int_\Omega \chi_A(u) \mathbf{B}^T \mathbf{CB} dV \tag{6.10}$$

As to a fuzzy element, it can be viewed as the extension of a conventional element and its definition domain is a fuzzy subset $\tilde{A}$ on Ω. A membership function $\mu_{\tilde{A}}(u)$ maps one point on Ω to a real number in the closed range $(0, 1)$, making the fuzzy element a super element defined on the fuzzy subset $\tilde{A}$. In this way, the stiffness matrix of a fuzzy element is of a similar form as that of a conventional element, while the only difference is that it is defined on a fuzzy subset as follows:

$$\tilde{\mathbf{k}} = \int_{\tilde{A}} \mathbf{B}^T \mathbf{CB} dV \tag{6.11}$$

The stiffness matrix of a fuzzy element in the above equation can be further expressed as:

$$\tilde{\mathbf{k}} = \int_{\tilde{A}} \mathbf{B}^T \mathbf{CB} dV = \int_{R_2} \mathbf{B}^T \mathbf{CB} \mu_{\tilde{A}}(u) dV \tag{6.12}$$

where, $\mu_{\tilde{A}}(u)$ is the full membership function of an element, u(x, y, z), for a fuzzy set $\tilde{A}$ and is given by:

$$\mu_{\tilde{A}}(u) \hat{=} \mu_{\tilde{A}}(x) \mu_{\tilde{A}}(y) \mu_{\tilde{A}}(z) \tag{6.13}$$

in which, $\mu_{\tilde{A}}(x), \mu_{\tilde{A}}(y)$ and $\mu_{\tilde{A}}(z)$ are the partial membership function in directions x, y and z, respectively.

When $\mu_{\tilde{A}}(u) = 1$, the fuzzy element degrades to a conventional element, that is,

$$\tilde{\mathbf{k}} = \mathbf{k} \tag{6.14}$$

When a fuzzy element is used to simulate a semi-infinite domain, equation (6.12) should be changed to:

$$\tilde{\mathbf{k}} = \int_{R_2} \left(\mathbf{B}^\infty \right)^T \mathbf{CB}^\infty \mu_{\tilde{A}}(u) dV \tag{6.15}$$

where: $\mathbf{B}^{\infty}$ = strain matrix of the corresponding generated element which simulates a semi-infinite domain

An eight-node isoparametric fuzzy element was constructed by Xie and Zhang, as shown in Fig. 6.3. Its stiffness matrix is of the form:

$$\tilde{\mathbf{k}} = \int_{\tilde{A}}\left(\mathbf{B}^{\infty}\right)^{T}\mathbf{C}\mathbf{B}^{\infty}\mu_{\tilde{A}}(u)dV$$
$$= \int_{-1}^{1}\int_{-1}^{1}\int_{0}^{\infty}\left(\mathbf{B}^{\infty}\right)^{T}\mathbf{C}\mathbf{B}^{\infty}|\mathbf{J}|\mu_{\tilde{A}}(\xi)\mu_{\tilde{A}}(\eta)\mu_{\tilde{A}}(\zeta)d\xi d\eta d\zeta \tag{6.16}$$

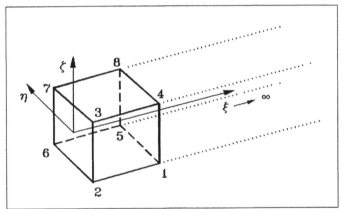

Figure 6.3 Eight-node isoparametric fuzzy element. (Xie and Zhang, 1985)

The comparison of the predicated displacements using either only conventional elements or the combination of conventional and fuzzy elements, indicates that the predictions by both implementations are close to the measured displacements. For the problem in their investigation, the implementation with fuzzy elements reduces the number of d.o.f. by 120 and the calculation time from 474 seconds to 245 seconds.

One remaining question to their study is whether the reduction of the number of d.o.f. can be achieved simply by using standard elements with larger physical size. If the answer is positive, then the necessity of using fuzzy elements will become minor.

Dynamic Analysis
An dynamic elasto-plastic constitutive model was proposed, as shown in Fig. 6.4. The failure surface is described by the following function:

$$f_1^* = f_1^*(I_1, \sqrt{J_2}, \dot{\varepsilon}) \tag{6.17a}$$

or

$$I_1 = A_0 + A_1\dot{\varepsilon} + A_2\sqrt{J_2} + A_3\dot{\varepsilon}\sqrt{J_2} + A_4\left(\sqrt{J_2}\right)^2 + A_5(\dot{\varepsilon})^2 \tag{6.17b}$$

The cap-type yield surface is described by:

$$f_2^* = f_2^*(I_1, \sqrt{J_2}, H, \dot{\varepsilon}) \tag{6.18a}$$

or

$$\frac{\left(\sqrt{J_2}\right)^2}{B^2} + \frac{\left(I_1 - C(\dot{\varepsilon})\right)^2}{\left(R(\dot{\varepsilon})\right)^2 B^2} = 1 \tag{6.18b}$$

All parameters in equations (6.17b) and (6.18b) can be determined by regression analysis from experimental data.

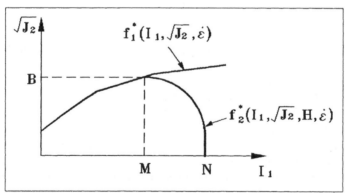

Figure 6.4 Yield and failure surfaces of an elasto-plastic model proposed by Xie and Zhang (1985).

The equilibrium equation for each element is of the form:

$$\left({}^t k_L + {}^t k_{NL}\right)\Delta u = {}^{t+\Delta t} r - {}^t q - m\,\Delta\ddot{u} \tag{6.19}$$

where: $'\mathbf{k}_L = \int_{'vol} \left('\mathbf{B}_L\right)^T {}'\mathbf{C} \, '\mathbf{B}_L \, dV$ = an element stiffness matrix corresponding

to the geometric linear strain components at instant t

$'\mathbf{k}_{NL} = \int_{'vol} \left('\mathbf{B}_{NL}\right)^T {}'\mathbf{C} \, '\mathbf{B}_{NL} \, dV$ = an element stiffness matrix

corresponding to the geometric non-linear strain components at instant t

$'\mathbf{q} = \int_{'vol} \left('\mathbf{B}_L\right)^T {}'\sigma \, dV$ = an internal force column matrix at time t

$^{t+\Delta t}\mathbf{r}$ = total external load column matrix

$'\mathbf{B}_L$ and $'\mathbf{B}_{NL}$ = geometric linear and non-linear strain matrices, respectively

$'\mathbf{C}$ and $'\sigma$ = constitutive matrix and Causy stress matrix, respectively;

$\Delta\mathbf{u}$ and $\Delta\ddot{\mathbf{u}}$ = column matrices of incremental displacement and acceleration at the element nodal points from time t to t+Δt, respectively.

$\mathbf{m}$ = element mass matrix

t and Δt = a time instant and time increment in a step-by-step integration scheme

$'vol$ = an element subdomain at instant t.

The wilson-θ method and a non-linear increment-iteration scheme were used in solving the nonlinear dynamic soil cutting problems. One problem associated to their study is that the maximum cutting speed is only 5 km/h. For the speed range below this value, the dynamic effect of tool draft is not fully developed. As a result, it is difficult to evaluate the validity of the model proposed by them.

6.1.4 Friction Element Analysis of Three-Dimensional Cutting (Liu and Hou)

Liu and Hou (1985) conducted three-dimensional FEM analysis of soil cutting using the friction element as interface elements. The detailed definition of friction element has been introduced in Section 4.2. Some other unique considerations are given below.

Soil Constitutive Model
The conventional approach to develop an elasto-plastic constitutive model is to choose a yield function and a work-hardening law first. Then to verify the correctness of the chosen yield function and make some modification on it. This

method cannot guarantee that the mathematical function is always consistent with experimental results. Huang (1983) proposed a new method to determine the yield function in a work-hardening law directly from laboratory test data. This avoids any man-made assumption and determines the plastic potential surface and its hardening parameters directly from soil stress-strain relation.

Based upon Huang's method, an elasto-plastic constitutive model for agricultural soil was proposed as illustrated in Fig. 6.5. AF is a shear failure line which is assumed to be independent upon stress history and a function of stress level. In a strain-hardening process, this line is assumed not to expand with the yield surface. The expression of AF line is of the form:

$$F_c = q_{oc} - CC - TF \times P_{oc} \qquad (6.20)$$

where, q_{oc} and P_{oc} are octahedral shear and normal stresses; CC and TF are test parameters.

The yield function was assumed to be a proportional ellipse surface. Based on the test data from a conventional triaxial apparatus, the following mathematical expression was obtained through data processing:

$$F_R = \left(\frac{P_{oc} - H_1}{K H_1} \right)^2 + \left(\frac{q_{oc}}{K R H_1} \right)^2 - 1 = 0 \qquad (6.21)$$

where, K and R are parameters determined from laboratory tests; H_1 is a hardening parameter of the form:

$$H_1 = \frac{M_5 P_a}{1 + K} \left[\exp \left(\frac{\varepsilon_v^p - M_3 \bar{\varepsilon}^p + \sqrt{\left(\varepsilon_v^p - M_3 \bar{\varepsilon}^p \right)^2 - 4 M_2 (\bar{\varepsilon}^p)^2}}{2 M_4} \right) - 1 \right] \qquad (6.22)$$

where: $\bar{\varepsilon}^p$ and ε_v^p = plastic strain components corresponding to octahedral shear and normal stresses

M_2, M_3, M_4 and M_5 = parameters determined from laboratory tests

P_a = atmosphere pressure

Progression of Soil Failure Surface Within the Soil Mass

In a soil cutting process, some shear failure or tension failure surfaces will be generated within the soil mass in front of a cutting tool. Previous investigators inserted joint or layer interface elements on a predefined failure surface. This artificially assumed that a failure surface existed at the beginning of a cutting process, but in reality the soil mass is an integrated continuum without any remarkable defects or crevices initially. Therefore, pre-inserting interface elements within the soil mass is not an approach with a sounding logic basis. Liu and Hou used the approach of controlling the state of elements to simulate the progression of soil shear or tension failure surfaces within the soil mass. In more details, the state of an element can be either dead or alive, which is in a similar fashion as in the ADINA program. The difference is that in ADINA the state of element is controlled by time period. While in Liu and Hou's study, the state is controlled by soil shear and/or tension failure criteria dependent upon element stress status. Whenever a soil element has failed due to shear or tension, the element stiffness is assigned a very small magnitude and is considered dead. This element will not contribute force resistance to the cutting tool in the following incremental loading steps. Thus, the discontinuities inside the soil mass can be simulated.

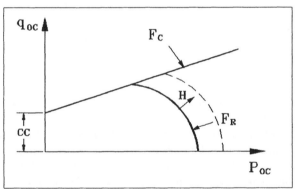

Figure 6.5 An elasto-plastic constitutive model proposed by Liu and Hou.

6.1.5 Three-Dimensional Analysis of Soil Failure with Curved Tools (Chi and Kushwaha)

Chi and Kushwaha (1990) presented a three-dimensional analysis of soil failure with flat and curved blades. Some special points are as follows.

Stress Correction
Most often, the tangent modulus on a stress-strain curve is evaluated from element stresses. In an incremental procedure, soil stresses initially increase with the blade displacement. When an element is failed due to shear or tension, a residual modulus with a small magnitude is usually assigned to this element. It

is possible to have a stress level outside of the failure surface. However, the soil stress should not fall outside the failure surface in a loading process. This incorrect stress level could affect the calculation results in the successive incremental steps. Therefore, in this situation the stress level should be corrected back to the failure surface. Davison and Chen (1974) presented a procedure to correct the stresses at constant mean stress for plane stress analysis. Nayak and Zienkiewicz (1972b) and Sirwardane (1980) proposed a procedure to correct the stresses normal to the yield surface. Nayak and Zienkiewitcz (1972a) transformed the Mohr-Coulomb criterion into a more convenient form using the stress invariant, as given by the following equations:

$$F = -\frac{I_1}{3}\sin\phi + \sqrt{J_2}\cos\psi_0 + \sqrt{J_2/3}\sin\phi - c\cos\phi$$
$$k_0 = 0$$
(6.23)

and

$$\sin\psi_0 = -\frac{3\sqrt{3}}{2}\frac{J_3}{\left(\sqrt{J_2}\right)^3}$$
(6.24)

where, F and k_0 are failure function and constant, respectively; I_1 is the first invariant of stress tensor; J_2 and J_3 are the second and third invariant of deviatoric stress tensors, respectively.

A positive value of the failure function in equation (6.23) means that current stress level is outside of the failure surface. In a stress correction process, the stress state is brought back to the failure surface in the normal direction. The corrected stress increment required to bring the stress level back to the failure surface is given by:

$$d\sigma = \frac{-F(^1\sigma)\dfrac{\partial F}{\partial \sigma}}{\left(\dfrac{\partial F}{\partial \sigma}\right)^T\dfrac{\partial F}{\partial \sigma}}$$
(6.25)

where: $^1\sigma$ = stress level before the stress correction

The corrected stress level is given by the following equation:

$$^2\sigma = {}^1\sigma + d\sigma$$
(6.26)

where: $^1\sigma$ and $^2\sigma$ = stress level before and after correction, respectively

Since the gradient is calculated based upon the stresses outside of the failure surface, it may not bring the corrected stresses exactly back to the failure surface. An iteration is used to repeat the above procedure until a satisfactory convergence is obtained.

Failure Zone
At each increment, normal, shear and principal stresses are calculated for each element. The stresses of each element are examined using the Mohr-Coulomb failure criterion. The failed elements are indicated by crosses in the plots. Thus, a gradual build-up of the failure zone is obtained by marking the failed elements in a plot for each incremental step. The analytical results show that initial soil failure started around the tool tip since the highest stress level usually occurs in this region. As the tool displacement increases, the failure zone expanded toward the soil surface. Finally a complete failure region is formed as shown in Figs. 6.6-6.9 for flat blades.

These failure zones form a curved failure surface in front of the tool, especially for the blade with 90° rake angle. The elements between the failure surface and the tool are still in a stable state. In a loading process, an inclined flat blade tended to push the soil forward and lift it up. The soil disturbance in the transverse direction is less for a inclined flat blade compared to a vertical one.

The failure zone obtained from the FEM analysis is not in agreement with Terzaghi's passive earth pressure theory in which it is assumed that all the soil in front of the tool up to the failure surface fails. However, it is observed that in a soil cutting test in a soil bin, some lumps of soil are left on the soil surface after an inclined blade passed. This may suggest that not all the soil in front of the blade failed. Therefore, it may be stated that the failure zone obtained from the FEA gives a closer approximation to the real situation in tillage, compared to using the Terzaghi theory.

The distance from the blade tip to the soil surface along the failure boundary increases with the rake angle. The blades with 30° and 45° rake angle show a minimum distance. This could give an explanation about why the draft force increases with rake angle significantly for the flat blade at the same cutting depth.

Normal Stress Distribution on Tool Surface
The normal stress contours are used to show the normal stress distribution on the surface of a tillage blade. The distributions at two typical rake angles are shown in Fig. 6.10. Large normal stresses are found along the tool edge. The

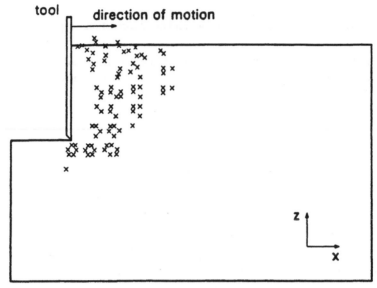

(a) forward and vertical failure

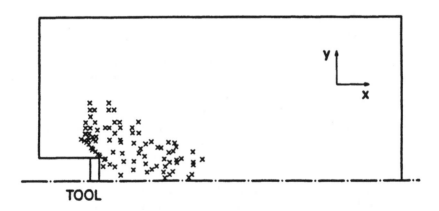

(b) horizontal failure

Figure 6.6 Failure zone of a vertical blade. (From *Journal of Terramechanics*, L. Chi et. al., 1990. Reproduction with permission of The International Society for Terrain Vehicle Systems, Hanover, NH.)

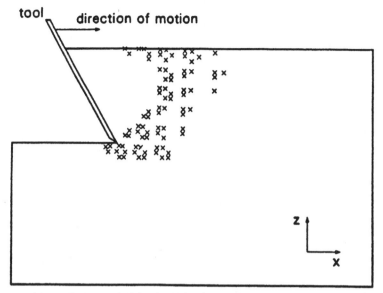

(a) forward and vertical failure

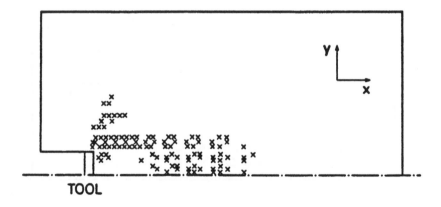

(b) horizontal failure

Figure 6.7 Failure zone of a blade with 60-degree rake angle. (Chi and Kushwaha, 1990.)

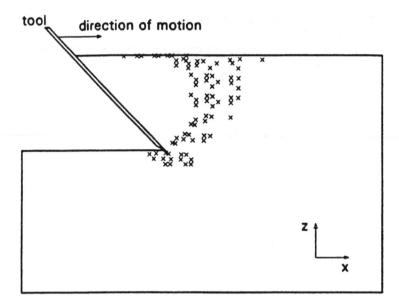

(a) forward and vertical failure

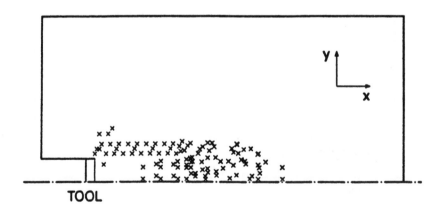

(b) horizontal failure

Figure 6.8 Failure zone of a blade with 45-degree rake angle. (Chi and Kushwaha, 1990.)

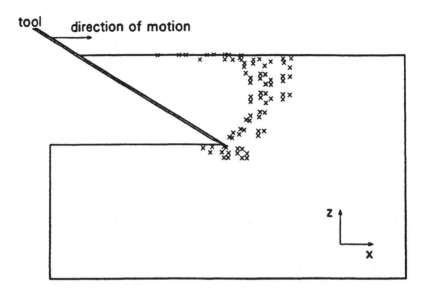

(a) forward and vertical failure

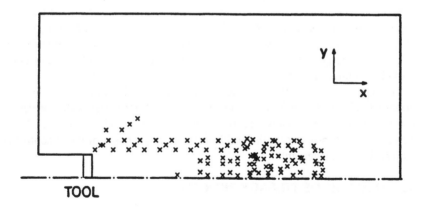

(b) horizontal failure

Figure 6.9 Failure zone of a blade with 30-degree rake angle. (Chi and Kushwaha, 1990.)

maximum stress occurs at the left and right bottom corners of the blade. Since the stress is highest, these two portions are expected suffering the greatest amount of wear. The magnitude of normal stress increases with rake angle, and the vertical tool position shows the highest stress level.

Curved Tillage Tools

The tool angle of a curved blade changes with depth. In order to simplify the problem, the circular curves are used for blade shape. Two sets of the FE analysis are conducted. In the first set of analysis, the tool angle at soil surface is selected as vertical ($90°$ rake angle) and the tool angles at bottom of the blade are given by 90, 60, 30 and zero degree, respectively. Fig. 6.11 shows the FE mesh of a curved blade with 90-degree tool angle at soil surface and 30-degree tool angle at the bottom. In second set of finite element analysis, the tool angle is set 30 degree at bottom and the tool angles at soil surface range from 30 (straight blade) to 90 degree (most curved blade).

The predicated drafts for different curvature are compared, as illustrated in Figs. 6.12 and 6.13. For a 90-degree tool angle at soil surface, an increase in curvature corresponds to a decrease in the tool angel at bottom. Therefore, the resistance to soil movement is reduced and lower drafts are obtained using curved blade, as illustrated in Fig. 6.12. This figure indicates that the minimum draft is obtained for a curved blade with a 30-degree tool angle at bottom. When the tool angle at bottom is fixed at 30 degree, an increase in curvature corresponds to an increase in the tool angle at soil surface. An increase in tool angle at soil surface results in an increase in soil resistance in front of the tool. In this situation, the draft is increased using curved blade, as illustrated in Fig. 6.13. The straight blade with tool angle of 30 degree at both top and bottom provides a minimum draft.

The results of the FE analysis show that the tool angle of curved blade plays a principal role in determining the extent of draft forces. Using the curved soil cutting blade do not always decrease the draft in all circumstances. It will, however, assist in inverting the soil block for the purposes of weed control and crop residue management.

6.2 ANALYSIS OF TILLAGE PROCESS

Tillage is a complex soil manipulating operation in crop production system. At present, there are, in general, three methods available to researchers to investigate tillage process. The first method is the field or indoor soil-bin experiments. The advantage of this method is that researchers can obtain first-hand experience on performance evaluation of different tillage tools. Since the physical conditions in fields are continuously changing, it is difficult to determine the basic rules governing soil-tool interaction from such experiments

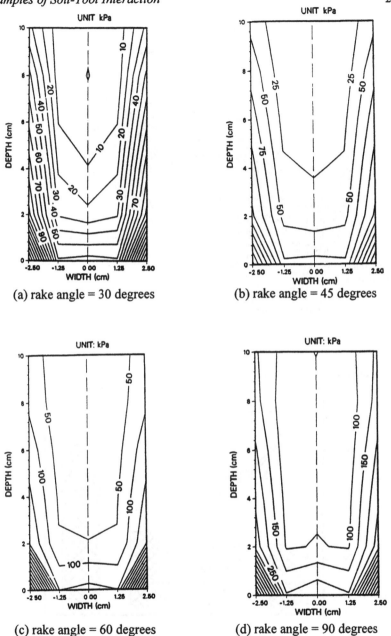

Figure 6.10 Normal stress distribution on the blade surface. (Chi and Kushwaha, 1990.)

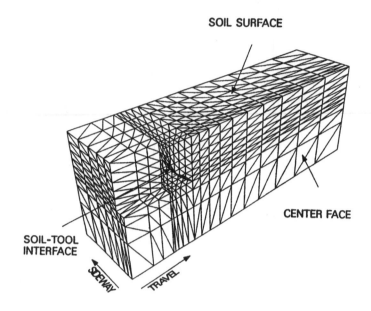

Figure 6.11 Finite element mesh with tool angle of 90 degrees at soil surface and 30 degrees at the bottom. (Chi and Kushwaha, 1990.)

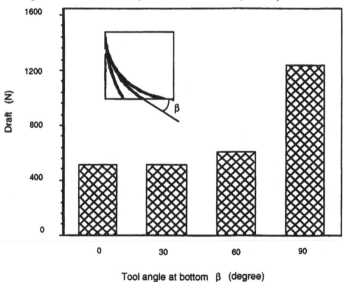

Figure 6.12 Draft with various tool angles for the blades with a fixed angle (90 degrees) at soil surface. (Chi and Kushwaha, 1990.)

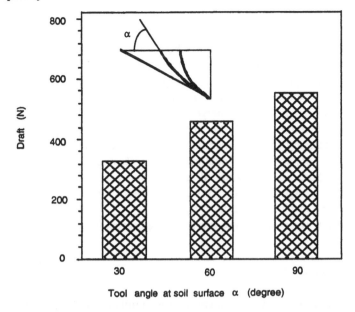

Figure 6.13 Draft with various tool angles for the blades with a fixed angle (30 degrees) at the bottom. (Chi and Kushwaha, 1990.)

conducted during a specific period of time and at a single location. If experiments are carried out at several locations for an extended period of times, the costs will be very high.

The second method is the analytical one which has been extensively used in the past for predicting draft under static or dynamic tillage conditions (Goodwin and Spoor, 1977; Hettiarachi and Reece, 1967; McKyes and Ali, 1977; Perumperal et al., 1983; Reece, 1965; Wismer and Luth, 1972; Zeng and Yao, 1992). The main advantage of analytical method is that it provides different types of simple formulations to idealize different relationships in tillage process. The designers of tillage tools can conduct some basic design activities according to the formulations provided by this method. However, most of these analytical models are based on conventional static or quasi-static equilibrium conditions which had the following deficiency:

(1) An assumed failure profile is a prerequisite to the draft prediction.
(2) The soil failure mode is affected by the tool speed. But, it is difficulty to define this effect using a changeable failure profile.
(3) Soil speed and acceleration area in front of tillage tool have to be simplified and assumed to follow a certain pattern.

The third method is numerical one including finite difference, finite element and boundary element methods (Yong and Hanna, 1977; Liu and Hou, 1985; Chi and Kushwaha, 1990; Wang and Gee-Clough, 1991; Xie and Zhang, 1985;

Zeng and Fu, 1985). By using numerical methods, the speed and acceleration fields of soil in front of a tillage tool can be calculated. These calculated values meet the exact requirement of the basic dynamic equation at every time interval point without any prerequisite about the soil failure profile. In addition, this method can deal with the interaction between complex-shaped tools and varying soil conditions compared to the analytical methods. The main disadvantages of the numerical methods include the requirement of certain expertise and he longer time necessary for implementing it. But, with the memory capacity and running speed of computers continuously increasing, the application of numerical methods in analyzing tillage process will generate a complete computer aided design procedure for tillage process similar to some successful cases in civil and mechanical engineering.

In this section, we will analyze the soil tillage process using finite element method.

6.2.1 Finite Element Model

Eight-node isoparametric elements are used to simulate soil dynamic response. The constitutive relation for this type of element is similar to that introduced in Section 6.2.2.1.

Thin-layer element is used to simulate soil-metal interface with the following element constitutive matrix:

$$[C^e]_I = \begin{bmatrix} C_{1I} & C_{2I} & C_{2I} & 0 & 0 & 0 \\ C_{2I} & C_{1I} & C_{2I} & 0 & 0 & 0 \\ C_{2I} & C_{2I} & C_{1I} & 0 & 0 & 0 \\ 0 & 0 & 0 & E_I & 0 & 0 \\ 0 & 0 & 0 & 0 & E_I & 0 \\ 0 & 0 & 0 & 0 & 0 & E_I \end{bmatrix} \tag{6.27}$$

and

$$C_{1I} = \frac{E_{ss}(1 - v_I)}{(1 + v_I)(1 - 2v_I)}$$

$$C_{2I} = \frac{E_{ss} \, v_{iI}}{(1 + v_{iI})(1 - 2v_{iI})}$$

where: v_i = Poisson's ratio of soil-metal interface element
         E_{ss} = initial tangent modulus of soil

E_i = tangent modulus of soil-metal interface element

E_i in the above equation is determined by:

$$E_t = K_t P_a \left[\frac{\sigma_n}{P_a}\right]^{n_i} \left[1 - \frac{R_{if}\, \tau}{\tau_{f0} + B_t\, \ln(\dot{\gamma}/\dot{\gamma}_0)]}\right]^2 \tag{6.28}$$

where: K_i, n_i = equation coefficients depending on soil-metal interface property

σ_n = normal stress at soil-metal interface

τ = actual shear stress at soil-metal interface

τ_{f0} = shear stress at soil failure obtained from direct shear-box test

R_{if} = failure ratio

$\dot{\gamma}$ = actual shear strain rate at soil-metal interface

$\dot{\gamma}_0$ = maximum strain rate on direct shear-box apparatus

B_i = coefficient relating to the sliding rate effect

The following differential equation is used for dynamic analysis of tillage:

$$\mathbf{M}\,\ddot{\mathbf{U}} + \mathbf{K}(\dot{\mathbf{U}}, \mathbf{U})\,\mathbf{U} + \mathbf{R} = 0$$

$$\ddot{\mathbf{U}} = \frac{d^2\mathbf{U}}{dt^2}; \qquad \dot{\mathbf{U}} = \frac{d\mathbf{U}}{dt} \tag{6.29}$$

where: $\mathbf{M}$, $\mathbf{K}$ = mass and stiffness matrix, respectively

$\mathbf{R}$ = external load vector

$\ddot{\mathbf{U}}$, $\dot{\mathbf{U}}$ and $\mathbf{U}$ = nodal acceleration, velocity, and displacement vectors, respectively

The soil parameters used in FEM calculation are listed in Table 6.1 with the idealized soil-cutting model shown in Fig. 6.14. The blade is considered a rigid object moving horizontally in soil at a depth of 0.05 m. The boundary conditions are given as follows:

1. The nodes on the bottom surface (KLM) are fixed in the vertical (Z) direction.
2. The nodes on the front and rear surface (AFIJLM and BCK) are fixed in the travel (X) direction.

3. The nodes on the symmetric surface (CKLJGD), which coincides with the center line of the moving tool, and left side surface ABM are fixed in sideways (Y) direction.
4. The nodes on the blade-soil interface (EDGH) have a specified displacement in travel direction during each increment.
5. The nodes on the side wall of the furrow (EHIF) are fixed in the sideways (Y) direction.
6. All other nodes are free in three directions.

Table 6.1 Soil parameters for constitutive models

Soil parameters:	
K_S	21.2
n_S	0.0
B_{sf}	0.093
cohesion (kPa)	6.0
int. Friction angle (degrees)	49.0
bulk density (kg/m³)	1070
Poisson's ratio	0.32
failure ratio	0.8
Soil-metal parameters:	
K_i	2.65
n_i	0.84
B_{if}	0.093
adhesion (kPa)	3.0
ext. Friction angle (degrees)	23.5
failure ratio	0.89

An incremental method is used to solve the soil non-linearity. The total horizontal movement of the flat blade is divided into many small increments. During each increment, a single-step standard Newton iteration method is used to solve equilibrium equations. The updated Lagrangian method is used to make continuous node coordinate transformation after each increment. Newmark step-by-step integration method is used to simulate the dynamic response of soil (Zienkiewicz and Taylor, 1989).

It is assumed that hydrostatic pressure did not cause compressive failure in agricultural soils. Only shear and tensile failures are considered with Mohr-Coulomb criterion and the tensile failure criterion of major principal stress.

Globally, soil tillage is a continuous soil cutting process in which soil and tool interact in the following continuous manner. At first, a certain part of the soil is activated from static state by the moving tool. Then this part of the soil enters the interaction stage with the tool. Finally the soil departs from the tool and the fresh part of the soil is activated. In this section, only one single sub-process is investigated which covers a period from the soil in a motionless state to the soil failure state.

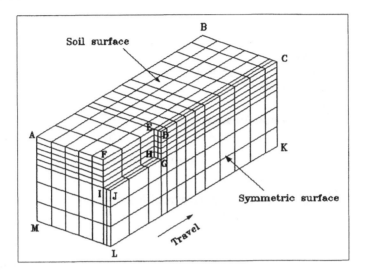

Figure 6.14 Finite element mesh.

6.2.2 ANALYTICAL RESULTS AND DISCUSSIONS

Development of Tool Draft
During the finite element analysis, the reaction forces are calculated from the summation of the node forces on the interface at each displacement increment. Fig. 6.15 shows the relation between draft and blade displacement at various travel speeds. It can be seen that at the lower speed range (less than 2.5 m/s), the draft increased gradually with the displacement of the blade and then tended to reach a peak value. For speeds above 2.5 m/s, lower speed curve first reached the peak value of draft, while the high-speed curve reached its maximum value after a longer movement of the blade. The probable reason for this phenomenon is that the high acceleration at high speed makes this effect last for a longer duration. The curve at 2.5 m/s indicates that the draft at higher speeds first increases quickly with the movement of the blade, then after reaching its peak, decreases gradually to the same level as the draft at lower speeds.

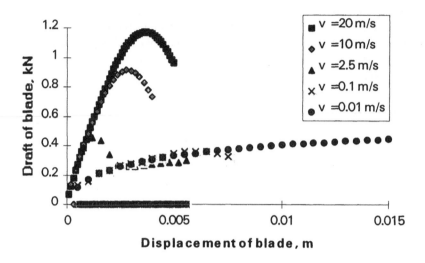

Figure 6.15 Blade draft vs. displacement at different speeds.

Development of Soil Failure Area

At each increment, normal stresses, shear stresses and principal stresses at each Gauss point are calculated and examined by the Mohr-Coulomb shear failure criterion and the tensile failure criterion of major principal stress. The shear-failure elements are marked by crosses, and the tensile-failure elements are indicated by circles.

The development of failure area at the speed of 0.01 m/s is shown in Fig. 6.16. At the beginning of the blade movement, only two small tensile areas exist beneath the blade and on the side wall of the furrow. when the blade moves a little further, soil shear failure occurs at some elements in front of the blade. Then the shear failure area will expand in front of the blade and also occurs both beneath the blade and on the side wall of the furrow near the blade. At this moment, two soil tensile areas also expand. At step (c), a complete shear failure area is formed in front of the blade and there is one tensile failure at the edge of this shear failure area which indicates that it approaches the failure point of soil body.

The development of failure area at the speed of 0.1 m/s is similar to that at 0.001 m/s. Fig. 6.17 indicates that for the speed of 2.5 m/s at the step (b) corresponding to the peak of draft, there exists only two tensile failure areas beneath the blade and on the side wall of the furrow with a very small shear failure spot in front of the blade. When the blade moves to past its peak draft, the shear failure area in front of the blade gradually increases in the same trend as the cases at the lower speeds, as shown in the step (c). The development of failure area at speeds of 20 m/s is similar to that at 2.5 m/s. The only difference is that a larger displacement is required to reach the peak draft at a speed of 20 m/s.

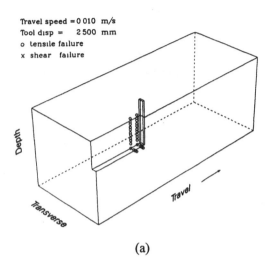

(a)

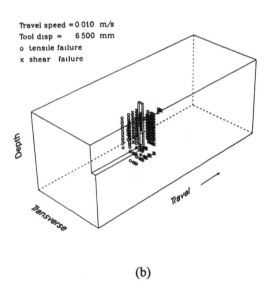

(b)

Figure 6.16 Development of soil failure area at 0.01 m/s.

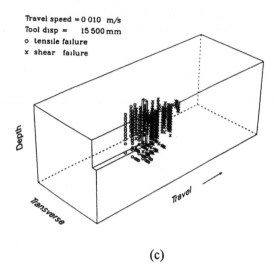

(c)

Figure 6.16 Continued.

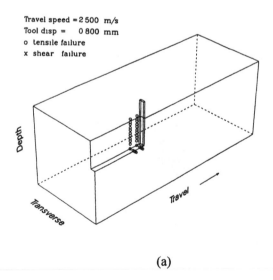

(a)

Figure 6.17 Development of soil failure area at 2.5 m/s.

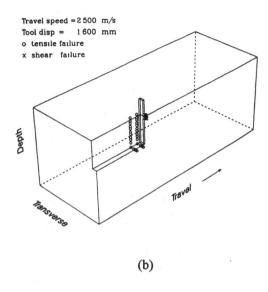

(b)

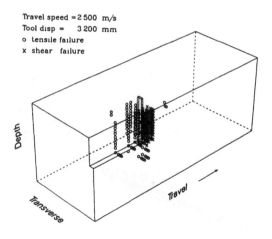

(c)

Figure 6.17 Continued.

Development of Soil Displacement Area

Fig. 6.18 shows the maximum displacement in soil vs. displacement of the blade. At lower speeds (0.01 and 0.1 m/s), the maximum displacement increases rapidly and deviates from the initial linear relation with the blade displacement before reaching the peak draft. But at speeds above 2.5 m/s, the maximum displacement maintains its linear relation with the blade displacement until the blade has moved a certain distance after passing its peak draft.

The soil movement pattern is displayed by plotting the nodal displacement vector, as shown in Fig. 6.19. The size of arrows is proportional to the magnitude of the displacement, and the direction of the arrow indicated the direction of the displacement. In order to demonstrate the displacement pattern clearly, the displacement at different speeds is multiplied by a different scale factor. A calibrated displacement and its corresponding vector are shown in each plot in order to allow readers to justify the actual value of each arrow.

According to Fig. 6.19, at lower speeds of 0.01 and 0.1 m/s, the shape of the displacement area maintains almost the same throughout the whole sub-process, and is similar to the shear failure area in front of the blade at the peak draft point. The disturbance area is quite large at these speeds. Above the speed of 2.5 m/s, the size of deformation area is reduced to only the area in front of the blade, and gradually increases a little with the displacement of the blade, as shown in Fig. 6.20.

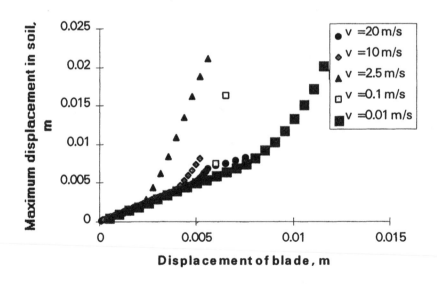

Figure 6.18 Maximum displacement in soil vs. displacement of the blade.

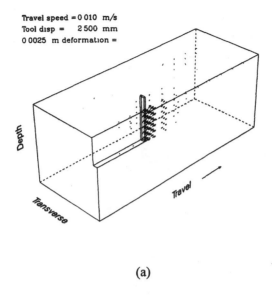

(a)

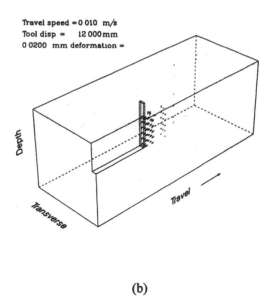

(b)

Figure 6.19 Development of soil displacement area at 0.01 m/s.

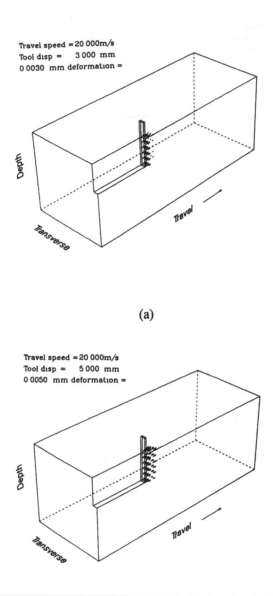

(a)

(b)

Figure 6.20 Development of soil displacement area at 20 m/s.

Development of Soil Speed Area

Fig. 6.21 shows maximum velocity in soil vs. displacement of the blade. At lower speeds (0.01 and 0.1 m/s), the maximum velocity in soil is negligible. At speeds above 2.5 m/s, initially, the maximum velocity is linear with the blade displacement, and then at a certain point before the peak draft point, deviates from the linear path and increases rapidly.

The soil speed area are also estimated by plotting the nodal speed vector. The treatment about the size and direction of the arrow is similar to the above. For the lower speeds (0.01 or 0.1 m/s), the size of speed area gradually decreases with the displacement of the blade, as shown in Fig. 6.22. At the speeds above 2.5 m/s, the speed area is reduced to the area in front of the blade, and the size of speed area gradually increases with the displacement of the blade until the peak draft point, as shown in Fig. 6.23.

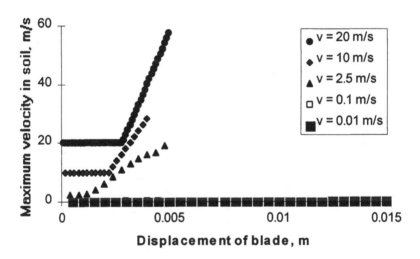

Figure 6.21 Maximum velocity in soil vs. displacement of blade.

Development of Soil Acceleration Area

Fig. 6.24 presents the relationship between maximum soil acceleration and displacement of the blade. The maximum acceleration in soil is negligible at the lower speeds (0.01 or 0.1 m/s). At speeds above 2.5 m/s, the maximum acceleration gradually increases with the displacement of the blade with a transition point corresponding approximately to the peak draft point.

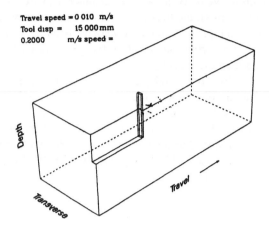

Figure 6.22 Typical soil speed area at 0.01 m/s.

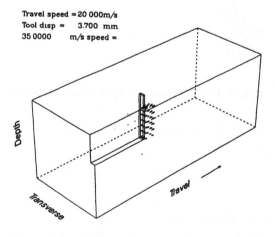

Figure 6.23 Typical soil speed area at 2.5 m/s.

For lower speeds (0.01 and 0.1 m/s), the acceleration arrow is forward in the beginning of blade movement, and then becomes backward. The size of acceleration area remains constant from the beginning of the blade movement until the peak draft point. Then the area gradually decreases in size (Fig. 6.25). At speeds above 2.5 m/s, the acceleration arrow is almost always forward. The change of its size is similar to that in lower speed cases, as shown in Fig. 6.26.

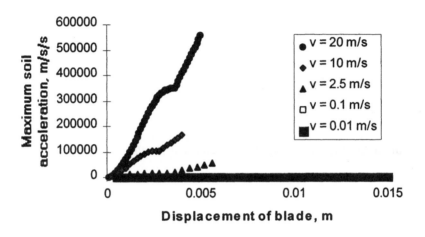

Figure 6.24 Maximum acceleration in soil vs. displacement of blade.

6.2.3 Summary

1. At lower speed ranges (less than 2.5 m/s), the draft increases gradually with the displacement of the blade and then tends to reach a peak value. For speeds above 2.5 m/s, the lower speed curve first reaches the peak value of draft, while the high-speed curve reaches its maximum value after a longer movement of the blade.

2. At lower speed range, there are only two tensile failure areas in the beginning and one shear failure is fully developed at the peak draft point; However, at the higher speed range (above 2.5 m/s), only two tensile failure areas exist from the beginning to the peak draft point.

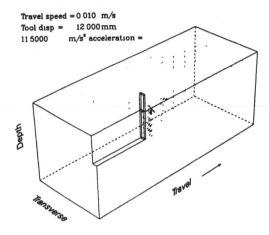

Figure 6.25 Typical soil acceleration area at 0.01 m/s.

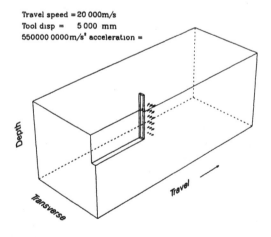

Figure 6.26 Typical soil acceleration area at 20 m/s.

3.   The shape of the displacement area remains almost the same throughout the whole sub-process at the lower speed range, and the disturbance area is considerably large at these speeds. Above the speed of 2.5 m/s, the size of deformation area is reduced to only the area in front of the blade.

4.   The velocity of soil is negligible at the lower speed range. At the higher speed range, the maximum velocity in soil, in the beginning, is equal to the speed of the blade, but at a certain point before the peak draft point, deviates from the linear path and increases rapidly.

5.   The acceleration in soil is negligible at the lower speed range. At speeds above 2.5 m/s, the acceleration area is reduced to the area in front of the blade, and the size of acceleration area gradually increases a little with the displacement of blade until the peak draft point.

6.3 EFFECT OF NUMERICAL PARAMETERS ON FINITE ELEMENT ANALYSIS

Soil tillage is a common operation in a crop production system. In the past, many investigators used FEM to analyze the soil-tool interaction during a tillage operation. However, in their studies, each investigator used his own numerical parameters in determining boundary conditions, mesh size, element type and soil failure criterion. Consequently, this influenced the comparability among the results of their studies. At present, there is practically no information available about the extent of the effect of these numerical parameters on the FEM results. The object of this section is to analyze the influence of the variation in boundary condition, mesh size, element type and soil failure criterion on the FEM results of the tool draft.

6.3.1 Idealized Interaction Between Soil and Tool

For a vertical tool moving horizontally in soil during tillage, the interaction between soil and tool is usually idealized to the patterns in Fig. 6.27 (a) and (b) for two-dimensional and three-dimensional cases, respectively. The numerical parameters of the idealization in Fig. 6.27 can be varied in several ways as described below.

Spatial Dimension
The tillage can be idealized to either a two-dimensional or a three-dimensional problem depending on the geometry of the tool. When the width of a tool is much greater than its cutting depth, the soil cutting can be approximately simplified to a plane strain problem. However, for most soil engaging tools, the ratio of width to depth is usually less than unity. Under this condition, the two-dimensional simplification may cause a noticeable error in the FEM prediction (Chi and Kushwaha, 1991).

Nonetheless, some investigators still used the two-dimensional FEM analysis because of its simplicity (Wang and Gee-Clough, 1991; Mojlaj et al. 1992). Fig. 6.28 shows the difference in draft prediction by using three-dimensional, plane strain and plane stress analyses. Here, the plane stress means that the thickness of soil elements is the same as that of the tool. According to Fig. 6.28, it can be inferred that the results of draft prediction by using plane strain and stress are almost the same, and noticeably less than that by using three-dimensional analysis that gives a closer prediction to the draft measured in the soil-bin.

Boundary Condition at Soil-Tool Interface
During a continuous soil cutting process, a certain relative displacement always exists at the soil-tool interface. The magnitude of this displacement depends on the roughness of the tool surface and the external friction and adhesion characteristics of the soil. Some investigators in their studies simplified the boundary condition at the soil-tool interface by considering the interface to be either a totally rough surface, i.e., no relative displacement existed, or a totally smooth surface, i.e., no friction existed , while other investigators used a kind of interface element such as joint element, friction element or thin-layer element to simulate this relative displacement phenomenon. Fig. 6.29 presents the difference of draft prediction under the three conditions listed below.
(1) Smooth state
   The interface is totally smooth where the nodes of soil elements at the soil-tool interface are free in depth direction for two-dimensional cases and in both depth and transverse directions for three-dimensional cases.
(2) Rough state
   The interface is totally rough where the nodes of soil elements at the soil-tool interface are fixed in depth direction for two-dimensional cases and in both depth and transverse directions for three-dimensional cases.
(3) Friction state
   There exists a certain relative displacement at the soil-tool interface where the nodes of soil elements next to the interface are connected to interface elements and the nodes of these interface elements at the interface are fixed in depth direction for two-dimensional cases and in both depth and transverse directions for three-dimensional cases.

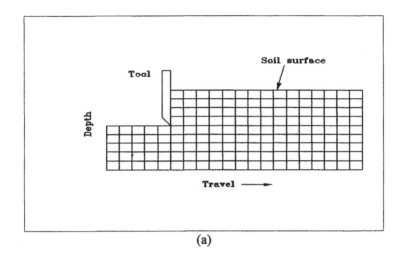

(a)

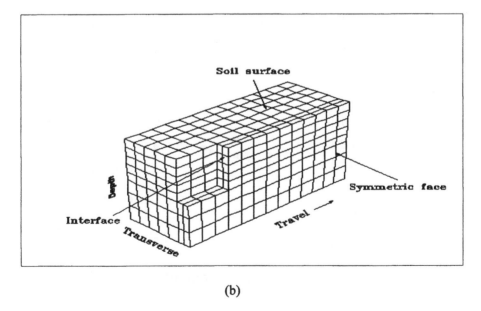

(b)

Figure 6.27 Spatial dimension of tillage problems.

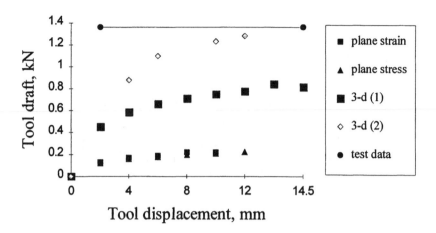

Figure 6.28 Difference in draft prediction by using different spatial dimensions.

Fig. 6.29 shows that the draft difference between the situation (1) and (3) is 7.2%, while the draft in situation (2) is noticeably larger than that in the former two situations.

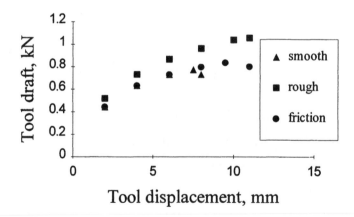

Figure 6.29 Draft prediction at different boundary conditions at soil-tool interface.

It is obvious that the reduction of friction at the interface will reduce the tool draft. However, by experimental method, it is difficult to comprehend the extent of maximum potential saving in tool draft for different tools and soil conditions by eliminating the friction at the interface. Using the finite element method, the theoretical maximum saving in tool draft can be estimated by the difference of draft prediction in situations (1) and (3). The maximum potential saving of the draft for a vertical blade is 7.2 % (Fig. 6.29). Zhang and Kushwaha (1993) reported the drafts for new and worn tillage tools (sweeps) as shown in Fig. 6.30. Due to the polishing effect of worn sweeps, the friction between a worn sweep and soil is lower than that between a new sweep and soil. The average draft reduction of the worn sweeps compared to the new sweeps is 10.3%. Although the tool and soil type in the above two cases are different, these data would provide a general concept about the influence of friction at the soil-tool interface on the tool draft.

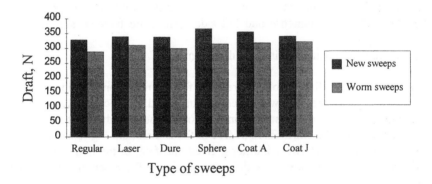

Figure 6.30 Measured drafts of new and worn sweeps. (Zhang and Kushwaha, 1993.)

Because soil cutting is a continuous process, an incremental method must be used to solve the material non-linearity of soil. The total horizontal movement of the tool is divided into many small increments. During each increment, a specified displacement in the travel direction of the tool is input as the main external load at the nodes of the interface for the models shown in Fig. 6.27.

Boundary Condition of Soil Body
In two-dimensional cases, the common definitions for the boundary conditions of the soil body are as follows:

(1) The nodes on the bottom boundary line are fixed in vertical (Y) direction

(2) The nodes on right and left boundary lines are fixed in horizontal (X) direction

(3) All other nodes are free in both X and Y directions

For the bottom line of the furrow, the nodes can be defined as either fixed or free in the Y direction. In addition, the same situation happens to the freedom in the Y direction at the intersecting point of the top surface line in front of the tool and the right boundary line. Fig. 6.31 indicates that the difference of draft prediction is negligible no matter whether the nodes on the bottom line of the furrow and the intersecting node of the top and right boundary lines are fixed or free in the Y direction.

In three-dimensional cases, the common definitions for the boundary conditions of soil body can be written as follows:

(1) The nodes on the bottom boundary surface are fixed in vertical (Z) direction
(2) The nodes on front and rear surface are fixed in travel (X) direction
(3) The nodes on symmetric and left side surface are fixed in sideways (Y) direction
(4) All other nodes are free in three directions

However, the definition of the freedom for the nodes on the side wall and bottom surface of the furrow is uncertain in Y and Z directions, respectively. The intersecting line of top surface in front of the tool and the front surface can also be defined to be either fixed or free in the Z direction. Fig. 6.31 shows the draft prediction under two conditions, fixed and free, at the nodes and in the directions mentioned above. The difference in the draft prediction appears to be negligible.

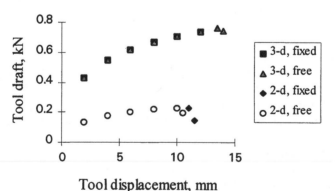

Figure 6.31 Draft prediction at different boundary conditions of soil body.

Mesh Density

Currently, there is little information available about how to construct a mesh model to simulate soil cutting. The main concern is how to make a good balance between running time cost and calculation accuracy. If a mesh is divided into a dense grid, the running time cost is not affordable, especially in three-dimensional cases. On the other hand, if a mesh is divided into a loose grid, the accuracy of the prediction may be questioned. Fig. 6.32 shows three mesh models for the same problem with different grid densities. Fig. 6.33 indicates that the predicted draft decreases with an increase in mesh density.

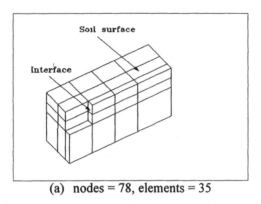

(a) nodes = 78, elements = 35

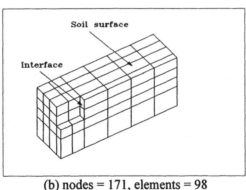

(b) nodes = 171, elements = 98

Figure 6.32 Different density of 3-D meshes.

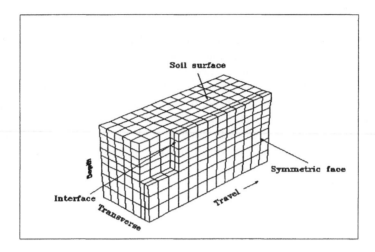

(c) nodes = 978, elements = 698

Figure 6.32   Continued.

6.3.2 Soil Elements

Soil is a very complex medium that possesses material non-linearity and geometry non-linearity during a continuous cutting process. There are several arbitrary features in defining the soil element as explained below.

Element Type
The most common elements used to simulate the soil cutting process in the past included the triangular and planar four-node isoparametric elements in two-dimensional cases, and eight-node isoparametric element in three-dimensional cases. Fig. 6.34 shows the difference of draft prediction by using triangular and planar four-node elements. The results indicated that the difference caused by the element types is negligible.

Soil Failure Criterion
The soil cutting process is a dynamic phenomenon and the soil confining pressure caused by a moving tool is much lower than that caused by a dam or a building. It is usually assumed that hydrostatic pressure did not cause compressive failure in agricultural soils. Only shear and tensile failures are considered in the FEM analysis.

Mohr-Coulomb criterion is the most suitable law to identify the state of soil failure. In detail, when half of the difference between major and minor principal stress $(\sigma_1 - \sigma_3)/2$ exceeds the shear strength at one element or one Gauss point, the shear failure is considered occurring at this element or point. Then , after soil failure, nonetheless there may still be a confining pressure existing at the element or Gauss point, the tangent modulus in shear stress-strain relation at the element or Gauss point should be reduced by a certain factor. Now, the question is how much the tangent modulus should be reduced after the shear failure. Fig. 6.35 shows the influence of different reduction factors in the initial tangent modulus on the draft prediction. The results indicate that the draft prediction remained nearly the same when the tangent modulus is reduced by a factor greater than 100. Therefore, in calculations, 1000 is selected as a suitable reduction factor for the initial tangent modulus after soil failure.

Soil Tension Criterion
According to the authors' knowledge, at present there is no information available about how to deal with soil tensile criterion in the study of soil cutting. In this paper, the tensile principal stress is used to identify the state of soil tensile failure. In detail, if σ_1 becames a tensile stress and is larger than $n \times c$ at one element or Gauss point, tensile failure is considered to occur. The tangent modulus is then modified to a small fraction of the initial tangent modulus. Here, 'c' stands for soil cohesion obtained from a triaxial apparatus and 'n' is a magnifying factor. Generally, the value of cohesion of soil samples obtained from the field is several times greater than that obtained from an indoor triaxial apparatus. Thus, n-fold cohesion is selected as a threshold to identify the tensile failure. Fig. 6.36 represents the draft prediction at different n-fold cohesion. The reduction factor after soil failure is chosen as 10^{-3}. In the range of 1-fold to 5-fold cohesion, the tool draft increased about 25.5 percent and varied almost linearly with n.

6.3.3 Dynamic Effect of Tillage

Damping Effect
The general matrix differential equation for time dependent problems can be expressed as follows:

$$\mathbf{M}\ddot{\mathbf{U}} + \mathbf{C}\dot{\mathbf{U}} + \mathbf{K}\mathbf{U} + \mathbf{R} = 0$$
$$\ddot{\mathbf{U}} = \frac{d^2\mathbf{U}}{dt^2}; \qquad \dot{\mathbf{U}} = \frac{d\mathbf{U}}{dt} \tag{6.30}$$

where: $\mathbf{M}, \mathbf{C}, \mathbf{K}$ = mass, damping and stiffness matrices, respectively

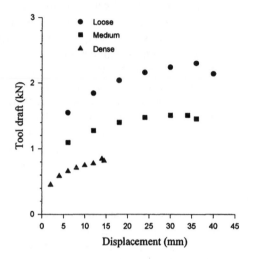

Figure 6.33   Draft prediction at different mesh densities.

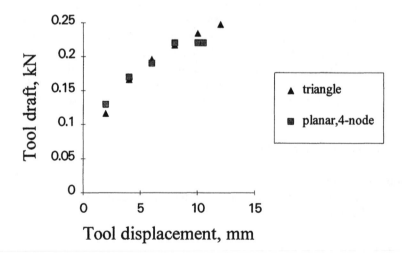

Figure 6.34   Draft prediction with different types of elements.

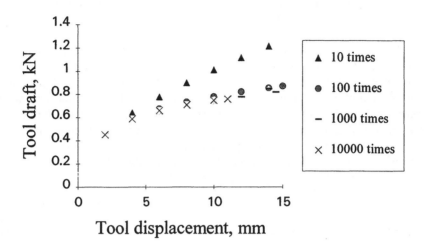

Figure 6.35 Draft prediction at different reduction factors of the initial tangent modulus after soil failure.

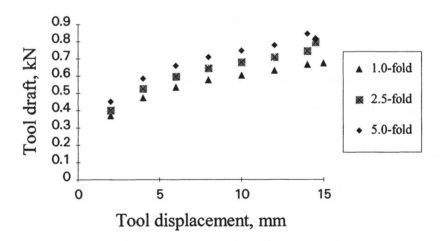

Figure 6.36 Draft prediction at different *n*-fold cohesion.

$\quad$ **R** = external load vector

$\ddot{\textbf{U}}$, $\dot{\textbf{U}}$ and **U** = nodal acceleration, velocity and displacement vectors, respectively

For soil-tool interactions, draft is determined by **R** column matrix, that, according to the above equation, can be considered to be comprised by three components: acceleration, damping and static equilibrium. Here, the contribution of acceleration to the tool draft is denoted by the product of mass matrix and acceleration vector; the contribution of damping effect is defined as the product of damping matrix and speed vector; and the contribution of static equilibrium is represented by the product of stiffness matrix and displacement vector. Because there is little information available about how to construct the damping matrix, the following equation is used for dynamic analysis of the soil-tool interaction as follows:

$$\textbf{M}\ \ddot{\textbf{U}} + \textbf{K}(\dot{\textbf{U}}, \textbf{U})\ \textbf{U} + \textbf{R} = 0 \qquad\qquad (6.31)$$

The damping matrix is neglected. However, the damping effect is moved into **K** matrix by considering **K** matrix to be a function of both displacement and speed vectors and by assuming the relation between **K** and $\dot{\textbf{U}}$ to be governed by the following two equations for soil elements and soil-metal interface elements, respectively.

The tangent modulus E_s of soil element is expressed as in equation (3.52):

$$E_s = E_{ss}\left[\frac{R_{sf}(1-\sin\phi)(\sigma_1-\sigma_3)}{2c\cos\phi+2\sigma_3\sin\phi+B_s(1-\sin\phi)\ln(\dot{\varepsilon}/\dot{\varepsilon}_0)}\right]^2 \quad \dot{\varepsilon} > \dot{\varepsilon}_0$$

$$E_s = E_{ss}\left[\frac{R_{sf}(1-\sin\phi)(\sigma_1-\sigma_3)}{2c\cos\phi+2\sigma_3\sin\phi}\right]^2 \qquad\qquad \dot{\varepsilon} > \dot{\varepsilon}_0$$

$$(3.52)$$

and

$$E_{ss} = K_s P_a\left(\frac{\sigma_3}{P_a}\right)^{n_s}$$

where: K_s, n_s = equation coefficients depending on soil property

$\quad\quad$ σ_1, σ_3 = major and minor principal stress in soil

$\quad\quad$ $(\sigma_1-\sigma_3)_{f0} = \sigma_1-\sigma_3$ at soil failure from conventional triaxial test

$\quad\quad$ P_a = atmosphere pressure

R_{sf} = failure ratio

$\dot{\varepsilon}$ = actual strain rate in soil

$\dot{\varepsilon}_0$ = base strain rate at conventional triaxial apparatus

B_s = coefficient relating to the strain rate effect

The tangent modulus E_i of soil-metal interface element is expressed by:

$$E_i = K_i P_a \left[\frac{\sigma_n}{P_a} \right]^{n_i} \left[1 - \frac{R_{if} \tau}{\tau_{f0} + B_i \ln(\dot{\gamma} / \dot{\gamma}_0)]} \right]^2 \tag{6.32}$$

where: K_i, n_i = equation coefficients depending on soil-metal interface property

σ_n = normal stress at soil-metal interface

τ = actual shear stress at soil-metal interface

τ_{f0} = shear stress at soil failure obtained from direct shear-box test

R_{if} = failure ratio

$\dot{\gamma}$ = actual shear strain rate at soil-metal interface

$\dot{\gamma}_0$ = maximum strain rate on direct shear-box apparatus

B_i = coefficient relating to the sliding rate effect

Fig. 6.37 shows the difference of draft prediction by choosing different B_s, while B_i is assumed to be equal to B_s. The maximum draft is approximately linear with the logarithm of B_s.

Dynamic Response
The step-by-step integration method can be used to simulate the dynamic response of soil. The time interval corresponds exactly to the load interval, and a very small value has to be chosen a very small value for high-speed cases. During each time step, acceleration at each node and velocity at each element or Gauss point are calculated, and soil tangent modulus at each Gauss point is updated continuously according to the present speed value and stress status. Fig. 6.38 demonstrates that the difference in draft prediction by using the Newmark and Wilson-θ methods is not remarkable.

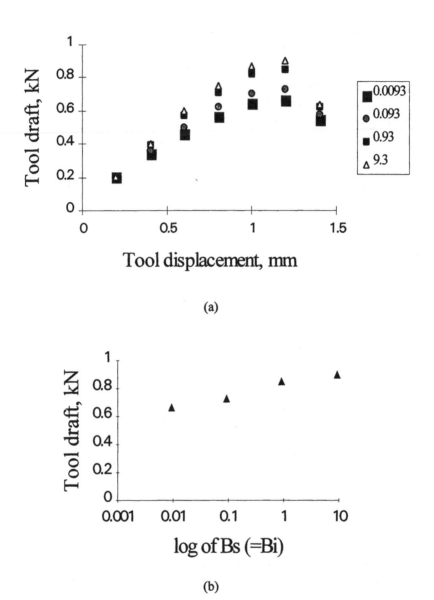

(a)

(b)

Figure 6.37   Draft Prediction at Different B_s and B_i.

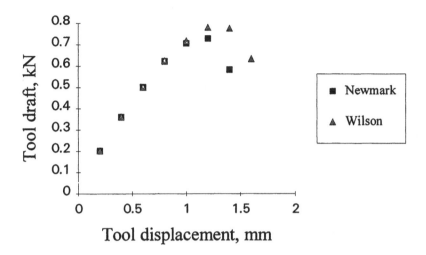

Figure 6.38 Draft prediction by Newmark and Wilson-θ methods.

6.3.4 Summary

1. The accuracy of draft prediction by using 2-dimensional plane strain and stress methods is almost the same for the soil-tool interactions using narrow tools. This accuracy is considerably less than that by using a 3-dimensional model.

2. Maximum potential savings in draft can be estimated by the difference between smooth and various friction states at the soil-tool interface. The variation in certain aspects related to the boundary condition of the soil body has little influence on prediction draft.

3. Mesh density has a noticeable influence on the draft prediction. The predicted draft decreases with an increase in the mesh density. The difference in draft prediction by using triangular and planar four-node elements is negligible.

4. Reducing the initial tangent modulus after soil failure by a factor greater than 100, the draft prediction remains nearly the same. A reduction factor of 1000 for the initial tangent modulus is found suitable after soil failure. Using the soil cohesion in the range of 1 to 5

times the laboratory test values, the tool draft increases about 25.5 percent and varies almost linearly with the multiplication factor.

5.   The predicted draft varies approximately linearly with the logarithm of the damping-effect coefficient B_s. The difference in draft prediction by using the Newmark and Wilson-θ methods is not significant.

7

PROGRAMMING TECHNIQUES FOR FINITE ELEMENT ANALYSES

In this chapter, we will introduce some concepts and techniques related to the implementation phase discussed in Section 5.1, which concerns the implementation details of computer languages for a finite element analysis.

7.1 PROGRAMMING LANGUAGES

Programming language is a main tool to realize algorithms for different finite element analyses. Generally speaking, there are two typical types of programming languages: imperative and functional languages. The von Neuman architecture causes the central features of the imperative languages to be:
(a) Variable. It models the memory cells;
(b) Assignment statement. It is based upon the piping operation;
(c) Iterative form of repetition. It is the most efficient method in this architecture, because the instructions are stored in adjacent cells of memory. But, it prevents the usage of recursion.

With a functional language, programming can be accomplished by functions to given parameters without variables, assignments, and iterations.

For scientific calculations like a finite element analysis, the imperative language is superior to the functional language. Among imperative languages, FORTRAN, PASCAL and C are good choices. Some features of these three languages are introduced below.

7.1.1 Introduction to FORTRAN Language

FORTRAN is the first compiled high-level language. In 1955, the original development environment of FORTRAN I was:

(a) IBM 704 was announced in 1954 with both indexing and floating-point instruction within the hardware. This ended the usage of interpretive languages as the main tool for scientific calculations;

(b) The characteristics of computers at that time included little memory, slow speed and unreliable performance;

(c) Scientific calculations were major applications;

(d) High cost of computers over programmers led to execution speed as the most important factor to consider in programming.

Under these circumstances, FORTRAN was developed by IBM mainly for scientific calculations. The history of FORTRAN is shown in Fig. 7.1 and explained as follows.

● FORTRAN I overview

FORTRAN I was introduced in 1957. All control statements were based upon IBM 704 instruction set. There were no data-type statements. Variables starting with I, J, K, L, M, N were implicitly integers, while others were implicitly designated floating-point numbers.

Its main advantage was that the machine code produced by the compiler was as efficient as what could be produced by hand.

● FORTRAN II overview

FORTRAN II was introduced one year after FORTRAN I. One of its new features was the independent compilation of subroutines. Without this feature, any change in a program required that the entire program be recompiled.

● FORTRAN IV overview

FORTRAN IV was introduced in 1962. The new features associated with this version include type declarations, a logic IF construct and the capability of passing parameters in calling subprograms to other called subprograms.

● FORTRAN 77 overview

FORTRAN 77 was launched in 1978 with new features: character string handling, logic loop control statement and IF statement with an optional ELSE clause.

● FORTRAN 8x overview

FORTRAN 8x possesses new capabilities of dynamically allocated arrays, many array operations, user-defined data type, modules and a CASE statement.

With the exception of FORTRAN 8x, the types and storage of all variables are fixed before run time. This sacrifices the flexibility to simplicity and efficiency, and eliminates the possibility of recursive subprograms and data structures that grow or change shape dynamically.

7.1.2 Introduction to PASCAL and C Language

PASCAL was announced in 1971. It was a descendent of ALGOL 68 and has had the largest impact on teaching programming. It was based upon its remarkable combination of simplicity and expressive power. Due to its feature of structured programming, although there are some insecurities in PASCAL, it is safer than FORTRAN.

The C language is another descendent of ALGOL 68. Although it was originally designed solely for system programming, it has been applied in many other fields due to its the capabilities: adequate control statements and data structures, a rich set of operators, inclusion in UNIX, and transportability among many computer systems. Its lack of complete type checking leads to both flexibility and insecurity.

The geneology of FORTRAN, PASCAL and C is shown in Fig. 7.1.

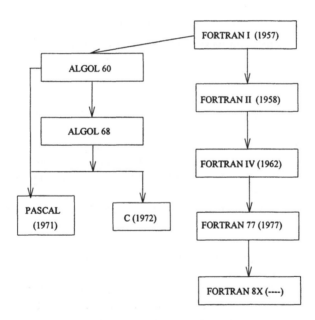

Figure 7.1 Geneology of FORTRAN, PACAL and C.

7.2   NOTATIONS FOR EVALUATING THE COST OF MEMORY STORAGE OR EXECUTION TIME

Before going deeply into the strategies for memory management and execution time reduction, it is necessary to introduce the notations that can be used to quantitatively evaluate the cost in space or time.

A function $f(n)$ is usually used to measure the space or time required to execute an algorithm or program on a problem of size n. Accordingly, we speak of the space cost or time cost functions of the algorithm or program. In general, the cost of executing a program increases with the problem size, n. If n is very small, the space cost is affordable and time cost is tolerable. With a big n, the space cost might become unaffordable and the program can no longer be executed in an acceptable period of time. Therefore, it is important to measure the space and time cost for large values of n. O-notation can be used to evaluate the asymptotic behavior of a cost function.

Definition: $f(n) = O(g(n))$ denotes " $f(n)$ has order at most $g(n)$ " and means that there exist positive constants C and N such that $|f(n)| \leq C\, g(n)$ for all $n > N$.

If we have two programs which execute a same task with the size of n, and the cost functions (either space or time) for the first and second are $O(n)$ and $O(n^2)$ respectively, it is easy to identify that the first program is superior to the second one in the sense of the cost. However, if the cost function for the first is changed to $O(10^4\, n)$, the answer is not straight. For $n < 10^4$, the second one is smaller, while for $n > 10^4$ the first one is smaller. Therefore, when using O-notation to compare two different programs, we should keep in the mind the actual range of problem size n. If n is assumed to be large enough, the following inequality is valid:

$$O(1) < O(\log n) < O(n) < O(n \log n) < O(n^2) < O(n^3) < \cdots < O(2^n)$$

where, $O(1)$ means that the cost is bounded by a constant. While the O-notation is used to express an upper bound, we might also wish to determine a lower bound of the cost function. Ω-notation can be used for this purpose.

Definition: $f(n) = \Omega(g(n))$ denotes " $f(n)$ has order at least $g(n)$ " and means that there exist positive constants C and N such that $|f(n)| \geq C\, g(n)$ for all $n > N$.

In some cases where the assertions $f(n)=O(g(n))$ and $f(n)=\Omega(g(n))$ are true at the same time, we will use the following notation:

Definition: $f(n) = \theta(g(n))$ denotes "$f(n)$ has the same order as $g(n)$" and means that there exist positive constants C_1, C_2 and N such that $C_1 g(n) \le |f(n)| \le C_2\, g(n)$ for all $n > N$.

$f(n) = \theta(g(n))$ means that $g(n)$ is both upper and lower bounds of the cost. This further infers that the costs at the worst and best cases take the same amount of space or time with only a variation of a constant factor.

7.3 STRATEGIES FOR MEMORY MANAGEMENT

Computer memory or storage is the place where program and data are located. In the prevalent von Neuman computer model as shown in Fig. 7.2, both program and data are stored in a single memory. Since the FEM applications in reality become more and more complex, how to allocate a limited memory space on a computer is a crucial factor to efficiently solve sophisticated problems.

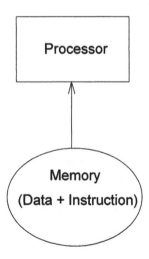

Figure 7.2 The von Neuman architecture.

7.3.1 Introduction to Memory Hierarchy

One primary aim of designing a memory system for modern computers is to provide programmers a large memory which can be accessed fast and with an optimally minimal cost. However, this goal is subject to restrictions of the following facts:

(a) Different types of memory have different cost per MByte. For instance, three main types of memory are SRAM (static random access memory), DRAM (dynamic random access memory) and magnetic disk. Their costs are $100-400, $25-50, and $1-2 per MByte in 1993, respectively.
(b) Different types of memory have different memory-access time. For instance, typical access time of SRAM, DRAM and magnetic disk are 8-35ns, 90-120ns and 10,000,000-20,000,000ns, respectively.
(c) Due to the different overheads for fetching a data from a given address, the access time of a small size of memory is faster than that of a large size of memory.

Since the fastest memories are more expensive per MByte than the slower memories, in order to be cost effective, the fastest memories are usually smaller. In addition, the observation of the principle of locality is also informative for a hardware designer to design a reasonable memory system. This principle states that programs access a relative portion of their address space at any instant of time. Two different types of locality exist:

(a) Temporal locality
If an memory address is accessed, it will tend to be accessed again soon. This refers to the locality in execution time.
(b) Spatial locality
If an memory address is accessed, it adjacent addresses will tend to be accessed soon. This refers to the locality in memory space.

A typical example of locality is the 90-10 rule, as shown in Fig. 7.3. This rule states that access within the top 10% and bottom 90% are uniformly distributed; that is, 90% of the execution time is spread evenly over 10% of a given program code and the other 10% of the execution time is spread evenly over the other 90% of the program code.
To make a balance among the differences in cost and access time, it is advantageous to build memory as a hierarchy of levels, with the faster memory closer the CPU (central processing unit) and the slower, less expensive memory below that, as shown in Fig. 7.4.

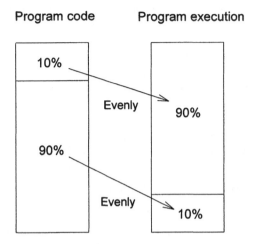

Figure 7.3 90-10 rule.

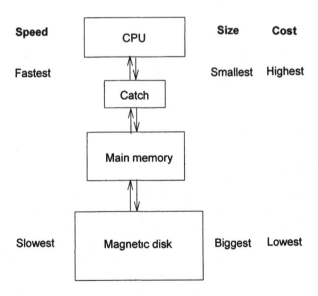

Figure 7.4 Memory hierarchy.

We take advantage of the principle of locality by letting the fastest and the most expansive memories be in a small size and hold the contents at the most frequently-accessed portion of addresses at any instant of time in executing a program.

As a detailed implementation shown in Fig. 7.4, main memory is implemented from DRAM, while caches can use SRAM. Cache is the level of the memory hierarchy between the CPU and main memory. The reasons to use a cache include:

(a) Due to the limitation of manufacturing technology and the requirement of high-speed data fetching, a processor is only allowed to have a limited number of registers to store data. For example, as to a RISC architecture, most often only 32 integer registers and 32 floating-point registers are available within the processor. This small number of registers cannot entirely accommodate any practical program.

(b) Even though the space of main memory is large enough to hold some practical programs, the access time is quite slow.

Therefore, to improve computer performance, a cache is inserted between the CPU and main memory. Whenever a computer is to fetch the content at a given address, it will first search the cache. If the cache contains the item, then it simply sends it to the processor through a data bus. Otherwise, accesses that fail inside the cache go to the lower memory hierarchy, main memory, which is larger and slower. According to the 90-10 rule, even though a cache is smaller, it can be responsible for almost 90% of execution, if it is large enough to hold 10% of a program code. Since a cache memory is usually at least 5 times faster than a main memory, the 90% of execution will be completed in much less time.

7.3.2 Allocation of Main Physical Memory

A main memory is the major place to hold an FEM program and to store intermediate calculation results. Before running a program, the entire code should be first loaded into the main memory. During the execution of the program, useful intermediate results should be kept in the main memory as much as possible rather than in the magnetic disks which are much slower.

For an FEM program, it should possess the capability of solving different sizes of problems. Thus, in a given case the program should automatically allocate working arrays, such as **K**, **U** and **R**, according to the upper bound of sizes of these arrays during execution. To accomplish this, there are two approaches available as follows.

Semidynamic Array Allocation

The concept of semidynamic array was first developed in ALGOL 60. A semidynamic array means that the subscript range is specified by variables, such that the size of the array is set at the moment storage is allocated to the array, which happens when the declaration is reached during execution.

In FORTRAN, this kind of array is called an adjustable array. An adjustable array is allowed to be declared only in a subroutine. The bounds of an adjustable array are transferred through subroutine arguments or a COMMON block from the upper-level module where the allocation of this adjustable array is specified.

Let us consider a simple 2-D FEM problem in which working arrays include only load matrix **RR**, displacement matrix **UU**, global stiffness matrix **KK**, nodal coordinate matrix **XC**, constraint condition matrix **UC**, and element information matrix **IX**. Different sizes of problems correspond to the different values of the following control information:

NJ -- number of nodes
NE -- number of elements
NU -- number of constraints
NF -- number of non-zero loads

These control information in turn determine the bounds of working arrays as:

UU (2, NJ) -- **UU** (1, i)= displacement in x direction at node i
$\qquad$ **UU** (2, i)= displacement in y direction at node i, where
$\qquad\qquad$ $1 \le i \le NJ$
XC (2, NJ) -- similar as above
UC (NU) -- It is assumed that each node has two freedoms. The
$\qquad$ number of global freedom corresponding to the i-th node
$\qquad$ is $2 \times i - 1$ and $2 \times i$.
RR (2, NF) -- When NF=0, i.e., no non-zero loads exist, this array
$\qquad$ becomes empty.
$\qquad$ **RR** (1, i) = number of global freedom corresponding to the
$\qquad\qquad$ i-th load
$\qquad$ **RR** (2, i) = magnitude of the i-th load, where $1 \le i \le NF$
IX (3, NE) -- **IX**(1, i), **IX**(2, i) and **IX**(3, i), respectively, corresponds to
$\qquad$ the nodal number of three nodes in i-th element, where
$\qquad$ $1 \le i \le NE$.
KK ($2 \times$ NJ, $2 \times$ NJ) -- It requires $(2 \times NJ)^2$ entries.

If only semidynamic arrays are used, it is not possible to allocate the above matrices individually in the main procedure. But, we can allocate all these arrays into a one-dimensional public array **M** which is declared in the main procedure. The criterion for declaring **M** array is that the size of **M** must be larger than the sum of all working arrays. Considering that no pointer data structure is available in most versions of FORTRAN, a set of integer variables are usually used as pointers specifying the starting position of each adjustable working array in **M** array. As an example of a detailed implementation of semidynamic array allocation, a program fragment is shown below. For convenience, //...// is used as comment delimiters which actually do not belong to FORTRAN.

C...............Main program

c...............Declare a public array M which is larger than the sum of sizes of
c...............all working
c...............arrays and set up a common block for control data

        COMMON/PUBLIC/M(30000)
        COMMON/CDATA/NJ, NE, NU, NF

c...............INPUT CONTROL DATA

        READ(5,*) NJ, NE, NU, NF

c...............Calculate the pointers

NUU = 1	//starting position for UU//
NXC = NUU+2*NJ	//starting position for XC//
NUC = NXC+2*NJ	//starting position for UC//
NRR = NUC+NU	//starting position for RR//
NIX= NRR+MAX(2, NF*2)	//starting position for IX//
NKK= NIX+3*NE	//starting position for KK//
NEND=NKK+4*NJ*NJ	

c...............Semidynamic allocation of working arrays through argument
c...............transfer between
c...............the main program and a subroutine or a calling and a callee
c...............procedures.

        CALL SUB1(M(NUU), M(NXC), M(NUC), M(NRR), M(NIX),
     *   M(NKK))

        END

C.................A subroutine

```
    SUBROUTINE SUB1(UU, XC, UC, RR, IX, KK)
    COMMON/CDATA/NJ, NE, NU, NF
    DIMENSION UU(*), XC(2, *), UC(*), RR(2, *), IX(3, *),
*                                    KK(2*NJ, *)
    ...
    END
```

It should be noted that the allocation space of **RR**(2, NF) depends solely on NF. If NF=0, then the array **RR** can not be allocated. To deal with this situation, the program allocates the space with a size of the maximum value between 2 and 2*NF to **RR**.

The basic idea of a semidynamic array allocation is that a series of working arrays with variable sizes is mapped, according to the calculated pointers, to a public array with a fixed large size. Although the size of the public array is not changeable, the pointers corresponding to the starting positions in the public array are allowed varying from one case to another. In this way, no space will be wasted in data storage and a maximum amount of space will be reserved to store the working arrays.

FORTRAN is a column-majored language, that is, when we map a two-dimensional array to its one dimensional counterpart, its row varies first followed by its column. For instance, $A(3,3)$ is mapped to $B(9)$ in the following way: $A(1,1)=B(1)$, $A(2, 1)=B(2)$, $A(3, 1)= B(3)$, $A(1, 2)=B(4)$, $A(2,2)=B(5)$, $A(3,3)=B(9)$. Due to the characteristic of column major, the following three expressions are equivalent in the above subroutine:

```
    DIMENSION XC(2, NJ)
    DIMENSION XC(2, *)
    DIMENSION XC(2, 1)
```

Dynamic Array Allocation
The concept of dynamic array was first developed in ALGOL 68. A dynamic array is one in which its declaration does not specify subscript bounds at all. In FORTRAN 8x, the storage space of an array can be allocated dynamically, that is, the bound of each dimension is set during execution rather than in compiling the source code.

In MICROSOFT FORTRAN 5.0, a dynamically allocated array must be declared that its property is ALLOCATABLE, and the variables for the bounds of each dimension should be specified. During execution, the ALLOCATE statement set the bounds of each dimension. When the array is not useful,

DEALLOCATE statement will release the storage space of the array to the common free space.

An example showing the usage of ALLOCATABLE property, ALLOCATE and DEALLOCATE statements is given below.

```
INTEGER        array1[ALLOCATABLE](: , : , : )
INTEGER        bound1, bound2, bound3
DATA           bound1, bound2, bound3, /10, 50, 100/
       ...
ALLOCATE (array1 (bound1, bound2, bound3), STAT = error)
       ...
DEALLOCATE  (array1, STAT = error)
```

7.3.3   Extension of Main Memory Space Using Virtual Memory Management

The storage space of main memory (or physical memory) is limited. Even though a 32-bit computer can support a potential physical address space up to 4 gigabytes, the practical settings of physical memories in most microcomputers and workstations are only 8 and 32 Megabytes, respectively. If user programs and their intermediate results exceed the space the physical memory can afford, the calculation will not continue or result in a wrong solution.

One former approach to solve this problem is to use the overlay technique by which it is the programmer's responsibility to divide programs into pieces and then to identify the pieces that are mutually exclusive. These mutually exclusive groups of pieces constitute overlays that are loaded and unloaded under the control of user program during execution. The programmer has to ensure that the program never tries to access an overlay not loaded, and the overlays loaded is not larger than the size of memory. The shortcoming associated to this approach is that controlling the overlay process adds a substantial burden for programmers.

The second approach is to develop new algorithms for solve FEM problems. For instance, as to the variable band and LDL^T method, its capacity for solving problems can be extended by blocking the stiffness and mass arrays (Mondkar and Powell, 1974). Then, only two of these blocks are necessary in RAM at any one time instead of the entire set of equations. However, this will increase I/O operations considerably and thus increase the time cost. An alternative to the blocked variable band method is frontal method. The frontal scheme works element by element, forming only the part of stiffness matrix belonging to the front. But the running speed is usually lower than the variable band method.

The third approach is to exploit the virtual memory. The concept of virtual memory was first implemented on the Atlas computer at the University of

Manchester. Its main idea is that the main memory can act as a 'cache' between processor and secondary storage such as magnetic disks. To implement the virtual memory technique, we need to:

(a) divide the main memory into pages each of which contains a certain amount of storage space.
(b) establish a virtual memory page table which translates a virtual address into a physical address.

Fig. 7.5 shows the principle to translate a virtual address into a physical address. The row entries of a page table is indexed with the virtual page number to obtain the corresponding portion of the physical address. The page table is usually located in the main memory with its starting address given by the page table pointer. The valid bit for each entry indicates whether the corresponding portion of the physical address is valid or not. If it is equal to 0, the searched page is in disk storage rather than in memory.

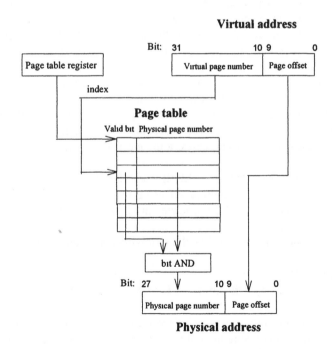

Figure 7.5 The translation from virtual address to physical address.

Since the length of virtual address is longer than that of physical address, if the virtual memory management is used, disk space can be viewed as a logical extension to physical memory. The paging hardware of processors can be used to map the virtual address to logical program addresses, allowing users' application programs of virtually unlimited size.

7.3.4 Case Study: Memory Management in DOS Environment

The DOS operation system was originally designed to operate on Intel86-family processors in 16-bit real mode, where a physical address space of 1 megabyte is supported. Since DOS allocates 384 Kilobytes for ROM and hardware subsystems, for users' program, there are only about 500 Kilobytes available which approximately corresponds to the size of a double precision array A(62000) and is a small address space for today's sophisticated applications.

The Intel 386 and 486 processors can operate in 32-bit protected mode. In such a mode, the physical address space is 4 gigabytes, and 32-bit operations are significantly faster than their 16-bit counterpart. Therefore, a 32-bit FORTRAN compiler can be chosen in which DOS extender places the processor in 32-bit protected mode. Since the current prices of Microcomputer RAM are in the order of $50.00 per megabyte (MB), the total RAM in 386 or 486 can be extended to 32 MB without much cost. In this way, the 640K memory barrier is broken.

In addition, if the code for virtual memory management (VM) in 32-bit FORTRAN compiler is chosen, the disk space can be viewed as a logical extension to physical memory, just like on minicomputers and mainframes. The paging hardware of the 386 and 486 processors can be used to map the virtual address to logical program addresses, allowing users' application programs of virtually unlimited size.

SVS FORTRAN software is a 32-bit compiler that can be run on a PC microcomputer, with the capabilities of using the DOS extender to place Intel 386 or 486 processors in a 32-bit protected mode, and implementing the virtual memory management.

To compile and link a source file, the following commands can be used:

      SVS   **filename.for**   **-c**   ↵   //only generate an .obj file//
      SVS   **file1.obj** **file2.obj** **-LLibf28** ↵   //only link .obj files//

where, bold characters refer to what a user needs to input through a keyword; ↵ denotes an action of pressing the return key one time. The convention of these two notations will be adopted throughout this section.

The useful filters which assist a compiling processing are listed below :

+q, =q, or -q: quiet +q, mostly quiet =q, and not quiet -q. Control the number
of messages generated during compilation.

-On: Set the optimization level from 0 to 4. The default is -04.

+vm or -vm: Bind/do not bind the virtual memory manager into the .exe file.
The default is +vm.

+COL72 or -COL72: Ignore/do not ignore source code past column 72. The
default is -COL72.

The procedures using the SVS compiler on a PC microcomputer are listed as
follows:

step1: **SVS_init** ⏎   //run the initializing file//

step2: **mkmf** ⏎   //generate a Make file//

step3: **edit   makefile** ⏎   //use the editor to edit the file with the name,
makefile

step4: in the editor window, insert a line with the content:
            $CFLAGS = + COL72$ //ignore source code past column 72//

step5: **make** ⏎

step6: **set** ⏎   //see the environment default//

step7: **modxconf   filename.exe** ⏎   //generate a swap file//

the default swapfile is c:\xmswap.tmp under the SVS C^3 extender.

7.3.5 Variable Band Storage of Global Stiffness Matrix K

Let us continue to use the example in Section 7.3.3. If the total number of nodes
is n and notations introduced in Section 7.2 are used, the space cost of matrix **K**
is $O(n^2)$, while the cost for each of other working arrays is $O(n)$. Thus, the
storage of matrix **K** should be paid particular attention.

If we store all entries of a large matrix **K** in memory , the cost for both space
and time will be enormous. This large storage space can be reduced by
developing a new scheme for storing the information contained in **K**, based
upon its intrinsic properties. The first property is that **K** is symmetric in most
cases. For a symmetric matrix, it is necessary to store only half of the whole
matrix, either a lower triangular or a upper triangular one. The second property
is that matrix **K** is sparse, that is, most entries in **K** are zeros. Storing and
calculating many entries with values of zero is an unnecessary waste. For
instance, if we have a **K** with 1000 rows ad columns, a full storage needs 10^6
memory cells. Fig. 7.6 shows a general configuration about the position of non-

zero entries in a stiffness matrix. Due to the symmetry of **K**, only the lower triangular part is shown.

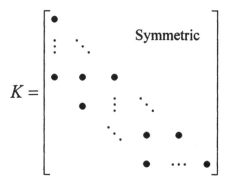

Figure 7.6 The position of non-zero entries, denoted by ●, in a matrix **K**.

If entries are numbered in the order of row by row, the shape of the group of non-zero entries in Fig. 7.6 shows a slanting band from the upper left to the lower right of the matrix. Suppose the width of this band is 100, then the total number of entries in this band is only 10^5, one tenth of the original figure. There are two common band storage methods. One is the constant band-width method, and the other the variant band-width method. The former is suited for a **K** with a regular profile of its non-zero band, while the latter can deal with a variety of irregular profiles of a band. The computation efficiency of the latter is slightly lower than the former in the case where the band profile is very regular. Most often, the variant band-width is chosen to store the global stiffness matrix due to its better overall performance.

To show how the variant band-width method works, a **K** with 6 rows and columns is given in Fig. 7.7. **K** has 36 entries in total, while there are only 13 entries enclosed by the dotted lines. It should be noted that the band width in each row is different, and zero entries outside the dotted lines do not need to be stored. However, zero entries inside the dotted lines still need to be saved. Thus, a one-dimensional array K1(13) can be declared to store the information contained in **K**. One problem arisen from transforming **K** into **K1**, is that one dimensional **K1** does not have the information about 2-dimensional location of K1(i) in **K**. To remedy this, another one dimensional array, **JP**, containing the number of storage order of the diagonal entries $K(i, i)$ in **K1** is needed. For the **K** in Fig. 7.7, there are six diagonal entries whose number of storage order in

K1 is 1, 3, 6, 8, 9 and 13, respectively. **JP**(6) is used to store these six integers and called the diagonal address array. With the help of the **JP** array, a one-to-one corresponding relation between **K** and **K1** can be established as follows:

$$K(i, j) = K1(JP(i) - i + j) \tag{7.1}$$

$$\mathbf{K} = \begin{bmatrix} 2.8 & & & & \text{Symmetric} & \\ 1.5 & 6.5 & & & & \\ 1.2 & 0 & 9.8 & & & \\ 0 & 0 & -1.3 & 7.1 & & \\ 0 & 0 & 0 & 0 & 1.5 & \\ 0 & 0 & 0.8 & 0 & 0 & 2.3 \end{bmatrix}$$

Figure 7.7 A profile of a band with a variant width

For instance, given i = 3 and j = 1, K(3, 1)=3.2, while **JP**(3)=6 and K(**JP**(3)-3+1)=K(4)=3.2. This supports the relation in equation (7.1). As to the diagonal entries in **K**, the above equation can be reduced to the form:

$$K(i, i) = K1(JP(i)) \tag{7.2}$$

In addition, the column number COL_i of the first non-zero entry in each row in **K** can be determined using **JP** as follows:

$$COL_i = i - JP(i) + JP(i - 1) + 1 \qquad (i \neq 1) \tag{7.3}$$

When i=1, COL_i is obviously equal to 1. Given i=5, COL_i=5-9+8+1=5 that means that in row 5, the entries in columns 1 through 4 are zeros.

For the small problem shown in Fig. 7.7, the total storage space required by **K1** and **JP** = 13+6 =19, while the entire lower triangle requires 21 memory cells. The saving is limited. However, for a **K** with 1000 rows ad columns, the variant band-width method will save tremendous memory space.

7.4   STRATEGIES FOR REDUCING THE TIME COST OF CALCULATION

One existing problem with finite element analyses is that it takes a long time to solve a problem compared to the traditional analytical method. Even on the mainframe computers, the calculating time is still very long because each user can only share a part of the total CPU time. This would discourage designers to carry out comparative designs repeatedly in a short period. This is the reason that designers have not adapted finite element method (FEM) in tillage tool designs. Hence, the calculating speed of the FEM analysis is a key factor in determining whether this method will be accepted by designers or not.

7.4.1 Background Information

The execution time of an FEM program depends upon many factors. Before discussing the detailed strategies for reducing the computation time, some concepts and background information are introduced in this subsection.

Influences of Computer Software and Hardware on the Execution Time
The influences of computer software on the execution time come mainly from the programming language and compiler a user chooses. Three important types of programming languages have been briefly introduced in Section 7.2. Among them, FORTRAN was designed primarily for scientific calculations and should be used as the first choice. The compiler technology has been continuously improved in the last three decades. Most of current compilers have the capability of optimization of a compiling process to generate an optimal machine code. For instance, with SVS FORTRAN compiler, four optimization levels are available. The higher the level, the longer the compiling process will be but the shorter the execution time of the machine code. In some cases, the difference in the execution time of a source code can be approximately about several times due to the combining influence of different compilers and different optimizing levels.

The influence of computer hardware is obvious. Different computers have different CPU designs, memory system designs, bus structures and I/O systems. The variation of these hardware configurations can lead to a different execution time for a same source code.

Table 7.1 shows the cost of time in two test programs under different hardware conditions. The first program is a test for the running speed of mathematical operations in which 15000×15000 multiplications are carried out. The time cost of a 486/50 microcomputer is longer than that of both SPARC station 2 and VAX 3100. The second program is an I/O type test in which two hundred records were written to and read from an unformatted and directly-accessed file for twenty times, with each record including one hundred integers. The time cost of this kind of I/O operation on a PC 486 was lower than that on both SPARC station 2 and VAX 3100.

Table 7.1 Cost of time in two test programs under different hardware conditions

Computer name	Status	Program 1 Time cost (m:s:s/100)	Program 2 Time cost (m:s:s/100)
PC 486/33	stand alone	5:04:23	0:11:53
PC 486/50	stand alone	3:23:94	/
SPACstation2	in network	2:51:65	0:15:99
VAX 3100 Model 90	in network	0:54:42	0:27:70

Performance Measures of an FEM Program
The execution time is the absolute measure of the performance of an FEM program. For two programs in the same software and hardware environment, the one taking less amount of execution time is faster. There are two types of time measurements available as follows:
(a) Elapsed time
   It refers to the latency to complete a program, including disk access, memory access, I/O, operating system, overhead, etc. This measure reflects the real length of time a user needs to take to complete the program, but get influences from many factors not related to the program.
(b) CPU time
   It refers to the latency to complete a program, excluding I/O or time in running other programs in a multi-task operating system. This measure can be further divided into two categories:

   - system CPU
      CPU time spent in the operating system.
   - user CPU
      CPU time spent in the user program.

The reason to pro CPU time measure is that I/O devices run independently of the CPU clock, i.e. asynchronously. Including I/O devices means that some random factors will be brought into the time measure. However, for a non-linear FEM analysis, some disk reads and writes may be necessary to store useful intermediate results in execution. In this case, we need consider the I/O operations.

To conduct a time measurement, we can use the time function provided in a programming language or an operating system. In Microsoft FORTRAN 5.0 for PC computers, the built-in subroutine GETTIM can be used, as illustrated in the following example:

```
        CALL GETTIM(ihr, imin, isec, i100h)
C..........The beginning of the portion of code to be measured
            • • •
C..........The end of the portion of code to be measured
        CALL  GETTIM(ihr2, imin2, isec2, i100h2)
        etime=(ihr2-ihr)*3600.0+(imin2-imin)*60.0+(isec2-isec)+(i100h2-
    *      i100h)/100.0
        WRITE (*, *) 'Execution time=', etime
        END
```

In Vax FORTRAN, the subroutine SECOND may be used, which is accurate to 0.01 second:

```
        REAL elapsed, t0, t1
        t0= 0.0
        t1=SECOND(t0)
C...........The beginning of the portion of code to be measured
            • • •
C...........The end of the portion of code to be measured
        elapsed= SECOND (t1)
        WRITE(*, 40) elapsed
40      FORMAT ('Execution time=', F12.6, 'second')
        END
```

In Sun FORTRAN, the subroutine DTIME can be used as follows:

```
        REAL   DTIME, TIMEDIFF, TIMEARRAY(2)
        TIMEDIFF=DTIME(TIMEARRAY)
C...........The beginning of the portion of code to be measured
            • • •
C...........The end of the portion of code to be measured
```

```
        TIMEDIFF=DTIME(TIMEARRAY)
        WRITE(6, 40) TIMEDIFF
40      FORMAT('Execution time=', F6.3, 'seconds')
        END
```

In Unix operating system, the <u>time</u> command returns:

$$<t_u> u \qquad\qquad <t_s> s \qquad\qquad <t_e> \qquad\qquad u\,\%$$

where: t_u = user CPU time

t_s = system CPU time

t_e = elapsed time

u = the percentage of user CPU time to elapsed time.

A simple timing program in C is given below:

```c
#include <stdio.h>
#include <time.h>
#include <sys/time.h>
extern clock_t clock();
#define CLK_TCK 1000000

main()
{
    struct tm *timer; int i,a;
    clock_t  start, finish, duration;
    start = clock( );

    for(i=1; I<100000; i++)
       a=i*i*i;        /* do multiplication */
    finish = clock( );
    printf( "process( ) took %f seconds to execute\n", ((double) (finish-
       start))/CLK_TCK);
}
```

7.4.2 General Criteria and Methods in Reducing the Execution Time

The following criteria have been extensively used to reduce the computation time:

Reducing More Complex Operations First
In a program, multiplication and division operations usually take longer time than addition and subtraction operations, although the detailed difference between these two kinds of operations varies among different computer architectures. The first effort should be made to reduce multiplication and division operations. The term, multiplication and division operations, will be abbreviated to m/d operations in the following text.

Lowering Array Dimension
Using high-dimensional array variables requires more time to search its elements than using simple variables or low-dimensional array variables. The variables should be used in the priority sequence of simple, low-dimensional array and high-dimensional array variables.

Using Unformatted Files in I/O Operations
Due to memory constraint, most intermediate calculation results need to be written to data files in the hard drive. These files should be opened in the unformatted form. A simulation result on a PC 486/33 indicates that the time cost of I/O operations of a formatted directly-accessed file was 2.3 times as long as that of the corresponding unformatted file.

The following schemes have been implemented in a recent compiler construction. The ideas behind these approaches can help the programmer to write a high-quality source code without relying upon the availability of a high-quality compiler.

Global/Local Common Subexpression Elimination
A transformation of a program is called local if it can be performed by looking only at the statements in a basic block; otherwise it is called global. An occurrence of an expression E is called a common subexpression if E was previously computed, and the values of variables in E have not changed since the previous computation. The process of a local common subexpression elimination is illustrated by the following example.

	Code before elimination	Code after elimination
1:	$S1 = 3*I$	$S1 = 3*I$
2:	$X = K1(S1)$	$X = K1(S1)$
3:	$S2 = 3*I$	
4:	$S3 = 3*J$	$S3 = 3*J$
5:	$S4 = K1(S3)$	$S4 = K1(S3)$
6:	$K1(S2) = S4$	$K1(S1) = S4$
7:	$S5 = 3*J$	
8:	$K1(S5) = X$	$K1(S3) = X$

Copy Propagation

The copy propagation means that a common subexpression can be substituted by a copy of the variable assigned by the original expression E. For instance,

Original code	Modified code
S1= A + B	S1 = A + B
S2= A + B	S2 = S1

where, the modified code save one addition. However, not in all cases, the copy propagation cannot implemented directly without any change, as shown in Fig. 7.8.

The reason for the failure shown in the above figure is that at the moment of calculating S3, the program do not know whether the left branch or right branch has been passed, and thus cannot determine whether S1 or S2 should be assigned to S3. The correctly-modified code is shown in Fig. 7.9, which saves on addition but causes an extra assignment.

Rearranging Statement Order

The purpose of rearranging the order of statements in a program is to keep the usage of a limited number of registers available in the processor to a possibly maximum extent. The average access time of a register is only 10 μs, while the access time of a cache and main memory is about 20 and 70 μs, respectively. Therefore, if more variables are allocated in registers, the program will be completed in a shorter time. But in a prevalent processor with RISC architecture, such as an MIPS processor, only 32 registers are available. Since some of these 32 registers have to be allocated solely for the operating system, we should use the remaining registers carefully.

Below, we give an example to demonstrate the influence of statement order in a high-level language on the compiling result in assembly language. For convenience, it is assumed that only two registers are available. The original source code and its compiling result in assembly language are shown in Fig. 7.10, while the source code modified by rearranging its order and the corresponding compiling result are presented in Fig. 7.11. Comparing two figures indicates that the compiling result changes from 10 to 8 instructions by rearranging the statement order.

Loop Optimizations

Three basic techniques are available as follows:

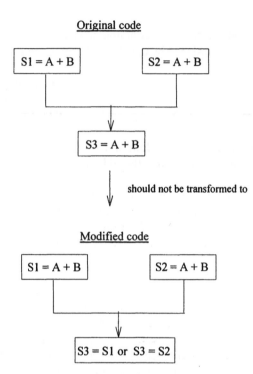

Figure 7.8 A failure of direct copy propagation.

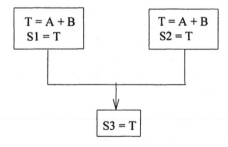

Figure 7.9 A correctly-modified code using copy propagation.

Original source code	Compiling result			
S1 = A+B	MOV	R1,	A	
	ADD	R1,	B	
S2 = C+D	MOV	R2,	C	
	ADD	R2,	D	
	MOV	S1,	R1	;store S1
S3 = E - S2	MOV	R1,	E	
	SUB	R1,	R2	
	MOV	R2,	S1	
S4 = S1-S3	SUB	R2,	R1	
	MOV	S4,	R2	

Figure 7.10 An original source code and its compiling result.

Modified code	Compiling result		
S2 = C+D	MOV	R1,	C
	ADD	R1,	D
S3 = E - S2	MOV	R2,	E
	SUB	R2,	R1
S1 = A+B	MOV	R1,	A
	ADD	R1,	B
S4 = S1-S3	SUB	R1,	R2
	MOV	S4,	R1

Figure 7.11 A modified code and its compiling result.

DO 200 I=1, LIMIT**4

$\downarrow$

T = LIMIT**4
DO 200 I=1, T

Figure 7.12 Moving an expression outside a loop.

(a) Code motion

This transformation moves an expression, which yields the same result independent of the number of times a loop is executed, from an inner loop to an outer loop or simply outside a loop. Fig. 7.12 illustrates an example of moving an expression outside a loop.

The benefit of code motion is obvious. Given a three-level loop with a size n in each level, one statement is the innermost level will take time $O(n^3)$. If it can be moved to the middle level or the outermost level, the time cost will be reduced to $O(n^2)$ or $O(n)$, repetitively.

(b) Induction-variable elimination

Induction variable is one whose value changes after each iteration. This kind of variable may be eliminated from a loop, as shown in Fig. 13, where an I.V. means an induction variable.

(c) Reduction in strength

This transformation means that in a loop an expansive operation should be replaced by a cheaper one whenever possible. For instance, a multiplication can be replaced by a cheaper addition or subtraction as illustrated in Figs 13, where an RIS denotes a place of reduction in strength.

	Original source code			Modified code
10:	I = M-1		10:	I = M-1
	J = N			J = N
	S1 = 6*N			S1 = 6*N
	V = K1(S1)			V = K1(S1)
				S2 = 6*I
				S4 = 6*J
20:	I = I+1	// I.V.//		
	S2 = I*6	//RIS//	20:	S2 = S2+6
	S3 = K1(S1)			S3 = K1(S1)
	IF(S3.LT.V) GOTO 20			IF(S3.LT.V) GOTO 20
30:	J = J-1	// I.V.//		
	S4 = 6*J	//RIS//	30:	S4 = S4 - 6
	S5 = K1(S4)			S5 = K1(S4)
	IF(S5.GT.V) GOTO 30			IF(S5.GT.V) GOTO 30
	• • •			• • •

Figure 7.13 Transformation of a source code using the induction variable elimination and reduction in strength.

7.4.3 Strategies in Reducing the Execution Time of an FEM Program

We enumerate typical calculations encountered in an FEM analysis as follows.

7.4.3.1 Shape Function and Inverse Jacobian Matrix
The shape function of 8-node 3-D isoparametric element is:

$$N_i = \frac{1}{8}(1 + \xi_i \xi)(1 + \eta_i \eta)(1 + \zeta_i \zeta) \qquad (i = 1, \cdots, 8) \tag{7.4}$$

where: N_i = shape function

Equation (7.4) can be changed to the following form in order to decrease one m/d operation:

$$N_i = (0.5 + S_i S)(0.5 + T_i T)(0.5 + Z_i Z) \tag{7.5}$$

where: $S_i = 0.5 \times \xi_i$ = a known constant

$S = \xi$

$T_i = 0.5 \times \eta_i$ = a known constant

$T = \eta$

$Z_i = 0.5 \times \zeta_i$ = a known constant

$Z = \zeta$

Similar changes can be made to the equation of $N_{i,\xi}, N_{i,\eta}$ and $N_{i,\zeta}$ to save another three m/d operations. Furthermore, by using mid-variables MS, MT and MZ in the following equation, the calculation amount of $N_{i,\xi}, N_{i,\eta}$ and $N_{i,\zeta}$ at one Gaussian point will be reduced from 16 m/d to 11 m/d operations.

$$MS = 0.5 + S_i S$$
$$MT = 0.5 + T_i T$$
$$MZ = 0.5 + Z_i Z$$
$$N_i = MS \times MT \times MZ$$
$$N_{i,\xi} = S_i \times MT \times MZ$$
$$N_{i,\eta} = MS \times T_i \times MZ$$
$$N_{i,\zeta} = MS \times MT \times Z_i \tag{7.6}$$

The inverse Jacobian matrix is directly derived as follows rather than by using the general subroutine to get inverse matrix which take many more m/d operations:

$$
\mathbf{J}^{-1} = \frac{1}{|\mathbf{J}|}
\begin{bmatrix}
Y_{,\eta}Z_{,\zeta} - Z_{,\eta}Y_{,\zeta} & Z_{,\xi}Y_{,\zeta} - Z_{,\zeta}Y_{,\xi} & Z_{,\eta}Y_{,\xi} - Z_{,\xi}Y_{,\eta} \\
X_{,\xi}Z_{,\eta} - Z_{,\zeta}X_{,\eta} & X_{,\xi}Z_{,\zeta} - Z_{,\xi}X_{,\zeta} & X_{,\eta}Z_{,\xi} - X_{,\xi}Z_{,\eta} \\
X_{,\eta}Y_{,\zeta} - Y_{,\eta}X_{,\zeta} & X_{,\zeta}Y_{,\xi} - X_{,\xi}Y_{,\zeta} & X_{,\xi}Y_{,\eta} - X_{,\eta}Y_{,\xi}
\end{bmatrix}
\tag{7.7}
$$

where: $|\mathbf{J}|$ = Jacobian determinant

$\mathbf{J}^{-1}$ = inverse Jacobian matrix

$X_{,j} = \partial X / \partial i, \quad i = \xi, \eta, \zeta$ (similar to other partial derivatives)

7.4.3.2 Element Stiffness Matrix
In 3-D cases, element strain submatrix $\mathbf{B_i}$, elasticity matrix $\mathbf{C}$ and element stiffness submatrix k_{ij} can be written as:

$$
\mathbf{C} =
\begin{bmatrix}
C_1 & C_2 & C_2 & 0 & 0 & 0 \\
C_2 & C_1 & C_2 & 0 & 0 & 0 \\
C_2 & C_2 & C_1 & 0 & 0 & 0 \\
0 & 0 & 0 & C_3 & 0 & 0 \\
0 & 0 & 0 & 0 & C_3 & 0 \\
0 & 0 & 0 & 0 & 0 & C_3
\end{bmatrix}
\tag{7.8}
$$

$$
\mathbf{B}_i =
\begin{bmatrix}
N_{i,x} & 0 & 0 \\
0 & N_{i,y} & 0 \\
0 & 0 & N_{i,z} \\
N_{i,y} & N_{i,x} & 0 \\
0 & N_{i,z} & N_{i,y} \\
N_{i,z} & 0 & N_{i,x}
\end{bmatrix}
\tag{7.9}
$$

$$
\mathbf{k}_{ij} = \int_{-1}^{1}\int_{-1}^{1}\int_{-1}^{1} \mathbf{B}_i^T \mathbf{C} \mathbf{B}_j^T |\mathbf{J}| d\xi d\eta d\zeta
\tag{7.10}
$$

where, $N_{i,x}, N_{i,y}$ and $N_{i,z}$ are derivatives of N to x, y and z respectively; v is Poisson's ratio; and

$$C_1 = \frac{E(1-v)}{(1+v)(1-2v)}$$

$$C_2 = \frac{Ev}{(1+v)(1-2v)}$$

$$C_3 = \frac{E(1+v)}{2}$$

If a common matrix operation is used, the calculation of $\mathbf{B}_i{}^T\mathbf{C}\mathbf{B}_j$ will need $3\times6\times6\times3 = 324$ m/d operations. By noticing that there are many zeros in both $\mathbf{B_i}$ and $\mathbf{C}$ matrices, $\mathbf{B}_i{}^T\mathbf{C}\mathbf{B}_j$ is written as follow:

$$\mathbf{B}_i{}^T\mathbf{C}\mathbf{B}_j =$$

$$
\begin{bmatrix}
\begin{array}{l}
N_{i,x}N_{j,x}C_1 + N_{i,y}N_{j,y}C_2 + N_{i,z}N_{j,z}C_3 \\
N_{i,y}N_{j,x}C_2 + N_{i,x}N_{j,y}C_3 \\
N_{i,z}N_{j,x}C_2 + N_{i,x}N_{j,z}C_3
\end{array}
&
\begin{array}{l}
N_{i,x}N_{j,y}C_2 + N_{i,y}N_{j,x}C_3 \\
N_{i,y}N_{j,y}C_1 + N_{i,x}N_{j,x}C_3 + N_{i,z}N_{j,z}C_3 \\
N_{i,z}N_{j,y}C_3
\end{array}
\\[3em]
&
\begin{array}{l}
N_{i,x}N_{j,z}C_2 + N_{i,z}N_{j,x}C_3 \\
N_{i,y}N_{j,z}C_2 + N_{i,z}N_{j,y}C_3 \\
N_{i,z}N_{j,z}C_1 + N_{i,y}N_{j,y}C_2 + N_{i,x}N_{j,x}C_3
\end{array}
\end{bmatrix} \quad (7.11)
$$

According to equation (7.11), the calculation of $\mathbf{B}_i{}^T\mathbf{C}\mathbf{B}_j$ only required 42 m/d operations. In addition, by referring to the symmetry of element stiffness matrix, only half of element stiffness matrix needed to be calculated. This further reduced the m/d operations to 21.

7.4.3.3 Solving the Global Stiffness Matrix K

Solving the equilibrium equations is the most time-consuming part in an FEM analysis, special attention should be paid. Any little improvement may result in a considerable amount of time saving.

If $\mathbf{LDL}^T$ method is used to solve the equations, there will be three layers of loops in the calculation. In the innermost loop, the execution times of any kind of calculation is proportional to n^3 (cubic of equation number) operations. Therefore, in the program, calculations should be tried to be performed in the outermost or middle loop which corresponded to $O(n)$ and $O(n^2)$ operations respectively.

Recall the triangular decomposition in Section 5.1 which uses the equation:

$$L_{ij} = k_{ij} - \sum_{k=1}^{j-1} \frac{L_{ik}L_{jk}}{L_{kk}} \qquad (i = 1,2,\cdots,n; \quad j = 1,\cdots,i) \tag{5.21}$$

The decomposition is actually a process of obtaining coefficients L_{ij}, which is carried out row by row from top to bottom. At each row, the calculation is proceeded column by column from left to right. No jump is allowed. The calculation of L_{ij} in equation (5.21) requires the values of entries only in the i-th and j-th rows as well as the diagonal entries above the j-th row. equation (5.21) involves j sets of calculation each of which contains 2 m/d operations. whenever a L_{jj} is obtained, it is stored in the form: $L_{jj}' = 1/L_{jj}$. Also, whenever all L_{ij}s in a row are solved, divide these elements by the diagonal element in this row, that is,

$$L_{jk}' = \frac{L_{jk}}{L_{kk}} = L_{jk} \times L_{kk}' \tag{7.12}$$

Since **K** is a sparse matrix, there may exist a range of zero entries in each row starting from the leftmost entry. According to equation (5.21), it is easy to verify that for k_{ij}s with zero value in such a range, the corresponding L_{ij}s are also equal to zero. Therefore, if we let COL_i and COL_j denote the column numbers of the first non-zero entry in i-th and j-th rows respectively, the multiplication of entries in the i-th row with those in the j-th row, may begin from the column whose column number is the larger one between COL_i and COL_j. If $COL_{ij} = \max(COL_i, COL_j)$, equation (5.21) can be transformed to:

$$L_{ij} = k_{ij} - \sum_{k=COL_{ij}}^{j-1} L_{ik}L_{jk}' \qquad (i = 1,2,\cdots,n; \quad j = COL_i,\cdots,i) \tag{7.13}$$

As to the forward elimination equation (5.24) which is rewritten as:

$$y_i = r_i - \sum_{j=1}^{i-1} \frac{L_{ij}y_j}{L_{ij}} \qquad (i = 1,2,\cdots,n), \tag{5.24}$$

since we store the diagonal elements in the form: $L_{jj}' = 1/L_{jj}$, and k_{ij}s, in the i-th row, whose column number is less that COL_i are zeros, equation (5.24) can be transformed to:

$$y_i = r_i - \sum_{j=COL_i}^{i-1} L_{ij}' y_j \qquad (i = 1,2,\cdots,n), \qquad (7.14)$$

The backward substitution equation (5.25) is rewritten as:

$$u_i = \frac{y_i - \sum_{k=i+1}^{n} L_{ki} u_k}{L_{ii}} \qquad (i = n, n-1, \cdots, 1) \qquad (5.25)$$

If we use the row-major variant-band-width storage, in backward substitution the following upper triangular equations need to be solved:

$$\mathbf{L}^T \mathbf{U} = \mathbf{Y} \qquad (5.22)$$

where the columns in $\mathbf{L}^T$ corresponds to the rows in $\mathbf{L}$. Since the calculation in equation (5.25) is proceeded on $\mathbf{L}^T$ row by row, that is, it is carried out on $\mathbf{L}$ column by column, this will not allow us to take advantage of the property of row-major variant-band-width storage. Thus, it is necessary to change the way of the backward substitution given in equation (5.25).

To simulate the process of a backward substitution on the upper triangular equations in equation (5.22), we first solve the last unknown variable u_n. Then, u_n is substituted into the $(n-1)$-th equation such that the u_{n-1} becomes solvable. Repeat this procedure until all unknown variables are solved. Note that the substitution of u_n into the first $(n-1)$ equations is proceeded column by column, that is, it is equivalently carried out on $\mathbf{L}$ row by row. This provides an opportunity to take advantage of the property of row-major variant-band-width storage.

As an example, let us consider a three-variable upper triangular equation:

$$\begin{aligned} L_{11} u_1 + L_{21} u_2 + L_{31} u_3 &= y_1 \\ L_{22} u_2 + L_{32} u_3 &= y_2 \\ L_{33} u_3 &= y_3 \end{aligned} \qquad (7.15)$$

From the third equation above, u_3 is obtained. Substitution of it into the first two equations and rearranging terms lead to:

$$L_{11} u_1 + L_{21} u_2 = y_1 - L_{31} u_3 = y_1'$$
$$L_{22} u_2 = y_2 - L_{32} u_3 = y_2'$$

From the second equation above, u_2 is solved. Substitution of it into the first equation and rearranging terms yield:

$$L_{11} u_1 = y_1' - L_{21} u_2 = y_1''$$

from which u_1 is obtainable.

Considering that equation coefficient L_{ij} has been transformed to $L_{ij}' = L_{ij}/L_{jj}$, we also need to transform the right-hand term into $y_i' = y_i/L_{ii} = y_i \times L_{ii}'$. Thus, equation (7.15) should be transformed to:

$$u_1 + L_{21}' u_2 + L_{31}' u_3 = y_1'$$
$$u_2 + L_{32}' u_3 = y_2' \qquad\qquad (7.16)$$
$$u_3 = y_3'$$

Therefore, the backward substitution can be carried out in the following form represented by an assignment statement:

$$u_j = y_j' - L_{ij}' u_i \quad (i = n, \cdots, 2; j = COL_i, \cdots, i - 1)$$

where: u_i = the variable that is already solved

7.4.3.4 Calculation of Element Strain and Stress

The following equations were derived to reduce 72 and 24 m/d operations from 144 and 36 m/d operations of an ordinary matrix multiplication for strain and stress calculation respectively.

$$
\mathbf{B}_i \mathbf{u}_i = \begin{bmatrix} N_{i,x} & 0 & 0 \\ 0 & N_{i,y} & 0 \\ 0 & 0 & N_{i,z} \\ N_{i,y} & N_{i,x} & 0 \\ 0 & N_{i,z} & N_{i,y} \\ N_{i,z} & 0 & N_{i,x} \end{bmatrix} \begin{bmatrix} u_i \\ v_i \\ w_i \end{bmatrix} = \begin{bmatrix} u_i N_{i,x} \\ v_i N_{i,y} \\ w_i N_{i,z} \\ u_i N_{i,y} + v_i N_{i,x} \\ v_i N_{i,z} + w_i N_{i,y} \\ u_i N_{i,z} + w_i N_{i,x} \end{bmatrix} \qquad (i = 1,8) \qquad (7.17)
$$

$$
\begin{bmatrix} d\sigma_{xx} \\ d\sigma_{yy} \\ d\sigma_{zz} \\ d\sigma_{xy} \\ d\sigma_{yz} \\ d\sigma_{zx} \end{bmatrix} = \begin{bmatrix} C_1 & C_2 & C_2 & 0 & 0 & 0 \\ C_2 & C_1 & C_2 & 0 & 0 & 0 \\ C_2 & C_2 & C_1 & 0 & 0 & 0 \\ 0 & 0 & 0 & C_3 & 0 & 0 \\ 0 & 0 & 0 & 0 & C_3 & 0 \\ 0 & 0 & 0 & 0 & 0 & C_3 \end{bmatrix} \begin{bmatrix} d\varepsilon_{xx} \\ d\varepsilon_{yy} \\ d\varepsilon_{zz} \\ d\varepsilon_{xy} \\ d\varepsilon_{yz} \\ d\varepsilon_{zx} \end{bmatrix}
$$

$$
= \begin{bmatrix} C_1 d\varepsilon_{xx} + C_2 d\varepsilon_{yy} + C_3 d\varepsilon_{zz} \\ C_2 d\varepsilon_{xx} + C_1 d\varepsilon_{yy} + C_2 d\varepsilon_{zz} \\ C_2 d\varepsilon_{xx} + C_2 d\varepsilon_{yy} + C_1 d\varepsilon_{zz} \\ C_3 d\varepsilon_{xy} \\ C_3 d\varepsilon_{yz} \\ C_3 d\varepsilon_{zx} \end{bmatrix}
$$

$$(7.18)$$

REFERENCES

Abdel-Hady, M. and Herrin, M. 1966. Characteristics of soil asphalt as a rate process. *Journal of the Highway Division, A.S.C.E.*, **92**(HW1):49-69.

Adachi, T. and Okamo, M. 1974. A constitutive equation for normality consolidated clay. *Soils and Foundations*, **14**(4):55-73.

Andersland, O.B. and Akili, W. 1967. Stress effect on creep rates of a frozen clay soil. *Geotechnique*. **17**:27-39.

Bailey, A.C., Johnson, C. E. and Schafer, R. L.. 1984. Hydrostatic compaction of agricultural soils. *Transactions of the Am. Soc. Agr. Engrs.* **27**(4):925-955.

Bathe, K.J., Ramm, E. and Wilson, E.L. 1975. Finite element formulations for large deformation dynamic analysis. *Intern. J. for Numerical Methods in Engineering.* **9**:353-386.

Bekker, M.G. 1960. *Introduction to Terrain-Vehicle Systems.* University of Michigan Press, Ann Arbor, Michigan.

Bekker, M.G. 1956. *Theory of Land Locomotion.* The University of Michigan , Ann Arbor, Michigan, 520p.

Bishop, A.W. 1966. The strength of soils as engineering materials. Sixth Rankine Lecture, *Geotechnique*, **16**(2):91-128.

Britto, A.M. and Gunn, M.J. 1987. *Critical State Soil Mechanics via Finite Elements.* Ellis Horwood Limited.

Chi, L. and Kushwaha, R.L. 1990. A non-linear 3-D finite element analysis of soil failure with tillage tools. *J. of Terramechanics.* **27**(4):343-366.

Chi, L. and Kushwaha, R.L. (in press). A non-linear three dimensional finite element analysis of soil failure with curved tillage tools. *Can. Agric. Eng.*

Chi, L. and Kushwaha, R.L. 1991. Finite element analysis of soil force on two tillage tools, *Can. Agric. Eng.*, **33**(1):39-45.

Chi, L. and Kushwaha, R.L. 1989. Finite element analysis of forces on a plane soil blade. *Can. Agric. Eng.* **31**(2):135-140.

Chi, L., Kushwaha, R.L. and Shen, J. 1993. An elasto-plastic constitutive model for agricultural cohesive soil. *Can. Agric. Eng.* **35**(4):245-251.

Christensen, R.W. and Wu, T.H. 1964. Analyses of clay deformation as a rate process. *J. Soil Mech. Foundations Div., Proc. A.S.C.E.*, **90**(6):125-157.

Christian, J.T. and Desai, C.S. 1977. Constitutive laws for geologic media. In: *Numerical Methods in Geotechnical Engineering.* (Desai, C.S. and Christian, J.T. eds.), McGraw-Hill Book Company.

Claar, P.W., Xie, L., Furleigh, D., Britton, R. and Inyang, H. 1995. Performance of hydraulic excavators for mining operations. *Proc. of 5th North American Conference/Workshop.* pp. 130-138.

Clough, G.W. and Duncan, J.M., 1971. Finite element analysis of retaining wall behavior, *J. Soil. Mech. Foundation Div., Proc. ASCE,* **97,** pp. 1657-1673.

Clough, G.W. and Duncan, J. M. 1969. *Finite Element Analysis of Port Allen and Old River Locks.* Report No. TE69-3, College of Engineering, Office of Research Service, Univ. of California, Berkeley, CA.

Davison, H.L. and Chen, W.F. 1974. *Elasto-plastic Large Deformation Response of Clay to Footing Loads.* Report No. 355-18, Dept. of Civil Engineering, Lehigh University, Bathlehem, PA.

Desai, C.S. and Siriwardane, H.J. 1984. *Constitutive Laws for Engineering Materials, with Emphasis on Geologic Materials.* Prentice-Hall, Inc., Englewood Cliffs, NJ.

Desai, C.S. and Phan, H.V. 1979. Three-dimensional finite element analysis including material and geometric nonlinearities. *Proceedings of the TICOM Second International Conference.* Oden, J.T. (ed). 205-224.

Desai, C.S., M.M, Zamin, J.G. Lightner and H.J. Siriwardane. 1984. Thin-layer element for interfaces and joints. International journal for numerical and analytical methods in geomechanics. **8:**19-43.

DiMaggio, F.L. and Sandler, I.S. 1971. Material model for granular soil. *J. of the Engineering Mechanics Division, ASCE.* **93**(3):935-950.

Drucker, D.C. and W. Prager, 1952. Soil mechanics and plastic analysis or limit design. *Q. Appl. Math.,* **10**(2):157-165.

Drucker, D.C., Gibson, R.E. and Henkel, D.J. 1955. Soil mechanics and work-hardening theories of plasticity. *Proceedings, ASCE,* Vol. **81,** Paper No. 798, pp.1-14.

Duncan, J.M. and Chang, C.Y. 1970. Nonlinear analysis of stress and strain in soils. *J. Soil Mech. Found. Div.,* Am. Soc. Civil Engrs. **96:**1629-1653.

Dunlap, W.H. and J.A. Weber. 1971. Compaction of an unsaturated soil under a general state of stress. *Transactions of ASAE.* **14**(4):601-607, 611.

Eyring, H. 1936. Viscosity, plasticity, and diffusion as examples of absoluter reaction rates. *Journal of chemical Physics.* **4**(4):283-291.

Eyring, H. and Halsey, G. 1948. *The Mechanical Properties of Textiles—The Simple Non-Newtonian Model.* High Polymer Physics, Chemical Rubber Co., Cleveland, OH, pp. 61-116.

Eyring, H. and Powell, R. 1944. Rheological properties of simple and colloidal systems. *Alexander's Colloid Chemistry.* **5:**236-252.

Goriatchkin, V. P. 1937. Kolesa zhatvennih mashin. *Sob Soch.*, Vol. II and V, Moscow.

Ghaboussi, J., Wilson, E.L. and Isenberg, J. 1973. Finite element for rock joints and interfaces. *Proc. Am. Soc. Civ. Eng. Soil Mechanics and Foundation.* **99**(SM10):833-848.

Gibbs, P. and Eyring, H. 1949. A theory for creep of ceramic bodies under constant load. *Canadian Journal of Science.* **27**:374-386.

Glasstone, S., Laidler, K. and Eyring, H. 1941. *The Theory of Rate Processes.* McGraw-Hill Book Co., Inc., New York, NY.

Godwin, R. J. and Spoor. G. 1977. Soil failure with narrow tines. *Journal of Agricultural Engineering Research.* **22**(4):213-228.

Goodman, R.E., Taylor, R.L. and Brekke, T. 1968. A model for the mechanism of jointed rock. *Proc. Am. Soc. Civ. Eng. Soil Mechanics and Foundation.* **94**(SM3):637-659.

Grevers, M.C.J. and Van Rees, K.C.J. 1995. Soil loosening of compacted forest landings in central Saskatchewan. *Proc. of 5th North American Conference/Workshop.* pp. 110-118.

Grisso, R. D. and Perumpral, J. V. 1985. Review of model for predicting performance of narrow tillage tool. *Transactions of the ASAE* **28**(4):1062-1067.

Herrin, M. and Jones, G. 1963. Behavior of bituminous materials from the viewpoint of absolute rate theory. Presented at meeting of the Amer. Assn. of Asphalt Paving Technologists, San Francisco, CA.

Hettiarachi, D. R. P.; Reece, A. R. 1967. Symmetrical three-dimensional soil failure. *Journal of Terramechanics.* **4**(3):45-67.

Hettiaratchi, D.R.P. and Reece, A.R. 1974. The calculation of passive soil resistance. *Geotechnique.*, **24**(3) 289-310.

Hettiaratchi, D.R.P., Witney, B.D. and Reece, A.R. 1966. The calculation of passive pressure in two-dimensional soil failure. *J. Agric. Engng. Res.*, **11**(2) 89-107.

Hill, R. 1950. *The Mathematical Theory of Plasticity.* Oxford University Press, London, England.

Horowitz, E. and Sahni, S. 1978. *Fundamentals of Computer Algorithms.* Computer Science Press, Potomac, MD.

Huang, W.X. 1983. *Soil Constitutive Modeling* (in Chinese). Hydraulic & Electric Power Press, Beijing, China.

Janbu, N. 1963. Soil compressibility as determined by oedometer and triaxial tests. *European Conference on Soil Mechanics and Foundation Engineering*, Wiesbaden, Germany, 1:19-25.

Keedwell, M. J. 1984. *Rheology and Soil Mechanics.* Elsevier Applied Science, London.

Koiter, W.T. 1953. Stress-strain relations, uniqueness and variational theorems for elastic-plastic materials with a singular yield surface. *Q. Appl. Math.* **11**:350.

Lade, P.V. 1977. Elasto-plastic stress-strain theory for cohesionless soil with curved yield surfaces. *Int. J. of Solids and Structure.* **13**:1019-1035.

Lade, P.V. and Duncan, J.M. 1975. Elastoplastic stress-strain theory for cohesionless soil. *J. of the Geotechnical Engineering Div., ASCE.* **101**(GT10):1037-1053.

Lade, P.V. and Nelson, R.B. 1984. Incrementalization procedure for elasto-plastic constitutive model with multiple intersecting yield surfaces. *Int. J. for Num. and Anal. Meth. in Geomechanics.* **8**:311-323.

Liu, Y. and Hou, Z.M. 1985. Three-dimensional nonlinear finite element analysis of soil cutting by narrow blades. In *Proc. of Int. Conf. on Soil Dynamics.* Auburn, AL. **2**:338-347. The National Tillage Machinery Laboratory, Auburn, AL.

Maximyak, R.V. 1968. Structural changes in clay soil due to deformation. In collective volume: Foundations and Underground Structures. Foundations and Underground Structures Reseach Institute., Moscow (in Russian), USSR.

McKyes, E. 1985. *Soil Cutting and Tillage.* Elsevier Science Publishing Company, New York, NY.

McKyes, E.; Ali, O. S. 1977. The cutting of soil by narrow blades. *Journal of Terramechanics.* **14**(2):43-58.

Mendelson, A. 1968. *Plasticity: Theory and Practice.* The Macmillan Company, New York, N.Y.

Mitchell, J.K., 1976. *Fundamentals of Soil Behavior.* John Wiley & Sons, Inc., New York, NY.

Mitchell, J.K., Campanella, R.G. and Singh, A. 1968. Soil creep as a rate process. *J. of Soil Mechanics and Foundation Division, ASCE,* **94**(SM1):231-253.

Mojlaj, E.G., Wulfson, D. and Adams, B.A., Analysis of soil cutting using critical state soil mechanics, *ASAE Winter meeting,* 1992, Paper No. 921501, Nashville, TN.

Murayama, S., and Shibata, T. 1958. On the rheological characteristic of clay. Part I, Bulletin No. 26, Disaster Prevention Research Institute, Kyoto University, Kyoto, Japan.

Nayak, G.C. and O.C. Zienkiewitcz. 1972. Convenient form of stress invariants for plasticity. *J. Struct. Div., Proc. of ASCE.* **98**(ST4):949-954.

Nayak, G.C. and Zienkiewicz, O.C. 1972. Elasto-plastic stress analysis. A generalization for various constitutive relations including strain softening. *Int. J. Numer. Meth. Eng.* **5**:113-135.

Nichols, M. L. and Read, I. F. 1934. Soil dynamics VI: Physical reactions of soils to mouldboard surfaces. *Agric. Eng.*, **15**:187.

Noble, C.A. and Demirel, T. 1969. Effect of temperature on the strength behavior of cohesive soil. *Highway Research Board Special Report 103*, pp. 204-219.

Nova, R. and Wood, D.M. 1979. A constitutive model for sand in triaxial compression. *Int. J. for Num. and Anal. Methods in Geomechanics.* **3**.

NRCC, 1986. *A Strategy for Tribology in Canada*, National Research Council Canada, Ottawa. NRCC 26556.

O'Callaghan J. R. and Farrelly, K. M. 1964. Cleavage of soil by tined implements. *Journal of Agricultural Research,* **9**:259-270.

Pan, J.-Z., Cai, G.-F. and Huang Y.-J. 1990. The modified rheological model for paddy soils in South China after remoulding. *J. of Terramechanics.* **27**(1): 1-6.

Payne, P. C. J. 1956. The relationship between the mechanical properties of soil and the performance of simple cultivation implements. *Journal of Agricultural Engineering Research,* **1**(1):23-50.

Payne, P. C. J. and Tanner, D. W. 1959. The relationship between rake angle and the performance of simple cultivation implements. *Journal of Agricultural Engineering Research*, **4**(4):312-325.

Pender, M.Z. 1977. A unified model for soil stress-strain behaviour. IXth ICSMFE, Tokyo. **2**:325-331.

Perumpral, J. V., Grisso, R. D. and Desai, C. S. 1983. A soil-tool model based on limit equilibrium analysis. *Transactions of the ASAE* **26**(4):991-995.

Polivka, M. and Best, C. 1960. *Investigation of the Problem of Creep in Concrete by Dorn's Method.* Univ. of California, Berkeley, CA.

Prager, W. 1959. *Introduction to Plasticity.* Addison-Wesley, Reading, MA.

Prandtl, L. 1920. Über die Härte plastischer Körper (The hardness of plastic bodies). *Nachr. kgl. Ges. Wiss.*, Gottingen, Math. Phys. Klesse.

Prevost, J.H. and Hoeg, K.K. 1975. Effective stress-strain strength model for soils. *J. of Geotechnical Engineering, ASCE,* **101**(GT3):259.

Ree, T. and Eyring, H. 1958. The relaxation theory of transport phenomena. In: *Rheology.* Vol. 2, Chapter 3, (F.R. Eirch, ed.), Academic Press, New York, NY.

Reece, A. R. 1965. The fundamental equation of earth-moving mechanics. Proceedings of the Symposium on Earth-moving Machinery *Mech. E.* **2**:8-14.

Roscoe, K.H. and Burland, J.B.. 1968. On the generalized stress-strain behavior of wet clay. In: *Engineering Plasticity*, (J. Heyman and F. Leckie, eds.) Cambridge University Press, Cambridge, England, pp. 535-609.

Sandler, I.S. and Baron, M.L. 1979. Recent development in the constitutive modeling of geological materials. *3rd Int. Conf. on Numerical Methods in Geomechanics*, Aachen.

Sandler, I.S., DiMaggio, F.L. and Baladi, G.Y. 1976. Generalized cap model for geologic materials. *J. of the Geotechnical Engineering Div., ASCE.* **102**(GT7):683-699.

Sebestra, R.W. 1989. Concepts of programming languages. The Benjamin/Cummings Publishing Company, Inc.

Shen, J. and Qun, Y. 1989. Investigation of creep characteristic of wet soils. *Proceedings of Int. Conf. of Appl. Mech. Vol.3*, Beijing, China.

Singh, A. and Mitchell, J.K. 1968. General stress-strain-time Function for Soils. *Journal of Soil Mech. and Found. Div., ASCE*, **94**:21-46.

Sirwardane, H.J. 1980. Nonlinear soil-structure interaction analysis of one-, two-, and three-dimensional problems using finite element method. Ph.D. dissertation, Virginia Polytechnic Institute and State University, Blacksburg, VA.

Swick, W. C. and Perumpral, J. V. 1988. A model for predicting soil-tool interaction. *Journal of Terramechanics* **25**(1):43-56.

Tanner, D.W. 1960. Further work on the relationship between rake angle and the performance of simple cultivation implements. *J. Agri. Engng. Res. 3*.

Terzaghi, K. 1959. *Theoretical Soil Mechanics*. J. Wiley & Sons Inc., N.Y.

Tremblay, J. and Bunt, R.B. 1979. *An Introduction to Computer Science. An Algorithmic Approach*. McGraw-Hill Book Company, New York, NY.

Vyalov, S. S., 1986. *Rheological Fundamentals of Soil Mechanics*, translated from the Russian by O.K. Sapunov. Elsevier, Amsterdam.

Wang, J. and Gee-Clough, D. 1991. Deformation and failure in wet clay soil. II. Simulation of Tine soil cutting. *Proc. of IAMC conference*, Beijing, China. Session **2**:219-226.

Wilde, P. 1979. *Mathematical and Physical Foundations of Elastoplastic Models for Granular Media*. Colloque Granco-Polonais, Paris.

Wilson, E.L. 1975. Finite elements for foundations, joints and fluids. In: *Finite Element in Geomechanics*. (ed. by G. Gudehus). 319-350. John Wiley & Sons, New York, NY.

Wismer, R. D. and Luth, H. J. 1972. Rate effect of soil cutting. *Journal of Terramechanics* **8**(3):11-21.

Wong, J.Y. 1993. *Theory of Ground Vehicles, Second Edition*, John Wiley, New York, NY.

Xie, X. 1983. Nonlinear finite element analysis of two-dimensional cutting problems in agricultural soils. *Transaction of Chinese Society of Agricultural Machinery*. (1):73-82.

Xie, X.M. and Zhang, D.J. 1985. An approach to 3-D nonlinear FE simulative method for investigation of soil-tool dynamic system. *Proc. of Int. Conf. on Soil Dynamics*. Auburn, AL. **2**:322-327.

Yao, Y. and Zeng, D. 1990. Investigation of the relationship between soil metal friction and sliding speed. *Journal of Terramechanics*. **27**(4):283-290.

Yong, R.N. and Hanna, A.W. 1977. Finite element analysis of plane soil cutting. *J. of Terramechanics*. **14**(3):103-125.

Yu, Q. and Shen, J. 1988. Research on rheological characteristic of wet soils (1), *Proc. of the 2nd Asian-Pacific Conference of the International Society for Terrain-Vehicle Systems*, pp. 23-39, Bangkok, Thailand.

Zeng, D. and Yao, Y. 1991. Investigation on the relationship between soil shear strength and shear strain rate. *Journal of Terramechanics*. **28**(1):1-10.

Zeng, D. and Yao, Y. 1992. A dynamic model for soil cutting by blade and tine. *J. of Terramechanics*. **29**(3):317-328.

Zeng, D. and Fu, Q. 1985. An approach to the analytical prediction in a rotary soil cutting process. *Proc. of Int. Conf. on Soil Dynamics*. Auburn, AL. **2**:428-442.

Zeng, D. and Yao, Y. 1992. A dynamic model for soil cutting by blade and tine. *J. Terramechanics*. **29**(3):317-327.

Zhang, J. and Kushwaha, R.L. 1993. Wear and draft of cultivator sweeps with hardened edges, *ASAE Summer Meeting*, Paper No. 931090.

Zhang, Q., Puri, V.M. and Manbeck, H.B. 1986. Determination of elastoplastic constitutive parameters for wheat en masse. *Transactions of the Am. Soc. Agr. Engrs.*, **29**(6):1739-1746.

Zienkiewicz, O. C. and Taylor, R. L. 1989. *The Finite Element Method*. McGraw-Hill Book Company, New York, NY.

Zienkiewicz, O.C., Best, B., Dullage, C. and Stagg, K.G. 1970. Analysis of non-linear problems in rock mechanics with particlular reference to jointed rock systems. *Proc. of the 2nd Congress of the international society for rock mechanics*. **3**:501-509. Belgrade, Yugoslavia.

Zienkiewicz, O.C. and Humpheson, C. 1977. Viscoplasticity: a generalized model for description of soil behavior. In: *Numerical Methods in Geotechnical Engineering*. (Desai, C.S. and Christian, J.T. eds.), McGraw-Hill Book Company, New York, NY.

INDEX

T - #0169 - 101024 - C0 - 229/152/19 [21] - CB - 9780824700812 - Gloss Lamination